Agriculture and
Natural Resources

Social Behavior and Natural Resources Series

Donald R. Field, Series Editor

Agriculture and Natural Resources

Planning for Educational Priorities for the Twenty-first Century

EDITED BY

Wava G. Haney
and Donald R. Field

Routledge
Taylor & Francis Group

LONDON AND NEW YORK

First published 1991 by Westview Press

Published 2018 by Routledge
52 Vanderbilt Avenue, New York, NY 10017
2 Park Square, Milton Park, Abingdon, Oxon OX14 4RN

Routledge is an imprint of the Taylor & Francis Group, an informa business

Library of Congress Cataloging-in-Publication Data
Agriculture and natural resources : planning for educational
    priorities for the twenty-first century / edited by Wava G. Haney
    and Donald R. Field.
      p. cm. — (Social behavior and natural resources series)
    Includes bibliographical references.
    ISBN 0-8133-8345-5
    1. Agriculture—Study and teaching (Higher)—Planning. 2. Natural resources—Study and teaching (Higher)—Planning. 3. Agriculture—Study and teaching (Higher)—Wisconsin—Planning. 4. Natural resources—Study and teaching (Higher)—Wisconsin—Planning. 5. Agricultural colleges—Planning. 6. University of Wisconsin System—Planning. 7. Strategic planning. 8. Strategic planning—Wisconsin. I. Haney, Wava G. II. Field, Donald R. III. Series.
S531.A57   1991
630'.71'1-dc20   91-14040
                                                  CIP

ISBN 13: 978-0-367-01243-4 (hbk)
ISBN 13: 978-0-367-16230-6 (pbk)

Contents

Foreword

In academic circles and private enterprise, strategic planning has become an effective tool to set goals, establish priorities, and prescribe actions. *Agriculture and Natural Resources: Planning for Educational Priorities in the Twenty-first Century* portrays both the process of planning and the substantive content driving a planning process. This collection of essays identifies issues and trends in agriculture, natural resources, and rural communities in the context of topical strategic planning. They suggest implications for programs, colleges, universities, and systems as they plan and budget for future instructional, research, and outreach programs.

It is not surprising that the University of Wisconsin System is a leader in strategic planning for agriculture and natural resources. Wisconsin's economic well-being has depended heavily on its natural resources, and the state has a long tradition of investing in its human resources through excellent schools and universities.

Wisconsin's three largest industries are agriculture, forestry, and tourism. Although dominated by dairying, its agricultural base is, however, quite diverse. Wisconsin is the top producer of vegetables for processing, including peas, snap beans, cabbage, beets, and cucumbers. In cranberry production, Wisconsin is second only to Massachusetts. Wisconsin's forests not only spur the state's paper industry, but result in an array of wood products that continue to expand in numbers and to increase in value of sales. Wisconsin's lakes and rivers, its rolling hills and sandstone bluffs, its wildlife and cultural events attract visitors to the state and encourage residents to travel throughout it.

The public universities in Wisconsin have a long tradition of developing and transferring technologies that help the state's industries to change, adapt, and grow. The System's land grant university, the University of Wisconsin–Madison, has a well-earned reputation for cutting-edge research. Its early scientists developed a milk test that revolutionized the dairy industry at the time and a procedure for adding vitamin D to milk to raise dietary standards. Their agricultural social scientists discovered what differentiates innovative businesses and developed a tradition of analyzing effects of agricultural policy on land tenure and land use. Today, UW–Madison's nutritional scientists cooperate with their medical school colleagues to discover relationships between diet and disease, and its biochemists and geneticists collaborate on advancing biotechnology.

Throughout this century, UW–Madison has trained some of the world's best agricultural scientists and policy makers. They are located in universities, businesses, and governments throughout the world. University of Wisconsin–Extension is recognized around the country for its rural community development and rural leadership programs. The University of Wisconsin–Stevens Point has one of the nation's largest undergraduate natural resource programs. The University of Wisconsin–River Falls, across the Mississippi River from Minneapolis and St. Paul, has developed a well-respected equestrian program. The University of Wisconsin–Platteville has revised its traditional agricultural curriculum to focus more on agribusiness, management, and farm profitability.

In the introductory chapter of this book, Ruth Carlson Robertson and Eugene P. Trani, the UW System academic vice president who has applied strategic planning to academic areas through a series of large-scale studies of several professional programs, discuss this concept and the process. The chapters that follow apply strategic planning to agriculture, forestry, recreation, and natural resource issues generally. They draw on the expertise of top people both from Wisconsin and around the country.

The studies are strategic in the sense that they concentrate on the long view and realistic educational needs, rather than on the all-too-familiar stopgap planning to which every one of us has been an unwilling party from time to time. The planning guidelines recommended by the studies support improved research and teaching environments by laying out developmental directions within which better individual and institutional strategies can be devised. They may also result in better overall long-term funding because of the clear sense of purpose they convey to state policy planners and lawmakers.

This volume is another example of planning for the future that has been a hallmark of the University of Wisconsin System. As president of that system, I am pleased to present a volume that I trust will help other universities with graduate and undergraduate programs in agriculture and natural resources to meet the challenge of the future.

Kenneth A. Shaw
President, the University of Wisconsin System

Acknowledgments

This book required the collective effort of a number of individuals who are concerned about agricultural and natural resource institutions, teaching, and research. These individuals represent a wide variety of organizations and institutions, come from many disciplines, and engage in a range of occupations and professions. Not surprisingly, they view agricultural and natural resource issues from diverse perspectives, use different types of analysis, and sometimes reach distinctive conclusions that contribute a richness to the discussion. The common threads are an overwhelming commitment to excellence and the centrality of agriculture and natural resource issues today and for the future.

We synthesized contributions from many people to compile this volume. Individuals from four different groups were important to this collaborative project: the authors of the chapters in Part II whom we invited as speakers and resource people to initiate our discussions at a systemwide conference; participants in the conference as well as the faculties of colleges of agriculture and natural resources in the state; UW System regents, administrators, academic, and professional staff; and members of the strategic planning committee. In particular, Chapters 2 and 11 reflect the collective wisdom and energy of the UW System Steering Committee for strategic planning for agriculture and natural resources. The committee members are listed at the end of this volume. We gratefully acknowledge their preparation of the draft material that formed the basis of these chapters.

We would not have entertained the idea of an edited volume without the encouragement of Vice President Eugene P. Trani, Associate Vice Presidents Albert J. Beaver and Dallas O. Peterson, and Acting Assistant Vice President Ruth Carlson Robertson. We would also like to thank Thomas L. Lyon, president of the UW Board of Regents, who provided advice and moral support for every phase of the committee's work and for this project. We would not have completed this volume without the editing and direct assistance of Gorden Hedahl. We also would have been lost without the staff assistance of Ginni Zuege and Paulette Napper in the preparation of the manuscript.

Wava G. Haney
Donald R. Field

I

Strategic Planning

1

Role of Topical Strategic Planning in University Systems

EUGENE P. TRANI
RUTH CARLSON ROBERTSON

Universities often feel besieged by demands for change. A growing and diverse clientele has an expansive list of research needs. Employers and students point out the need for curricular adjustments in response to an expanding knowledge base and a less than adequate level of basic and integrative skills. Recruitment and retention of faculty and staff is difficult in shrinking pools and necessitates expensive research packages to attract and retain the most promising. Research and instruction costs continue to rise, but higher education must compete for limited resources with other important state and national programs. How does a university system plan, establish priorities, and choose among alternative program activities under conditions of fiscal constraint? In the University of Wisconsin System, we have begun to meet these challenges through a series of topical strategic plans.

Strategic planning as outlined in this chapter emphasizes process. It is based on the assumption that involvement in an open process of gathering information, articulating a vision of the future, and establishing the goals and strategies to meet future challenges generates commitment and improves decision making.

Outcomes, of course, are important. They can send important messages to colleagues in the university and to the citizens of the state. Outcomes provide guidelines within which programs can be planned and budgets built. The strategic planning document can provide a concise overview of current conditions and anticipated future influences for the topic studied.

Inevitably, plans are limited by changes that may occur within the university, and more particularly, those likely to occur outside the university. The best futurists and planning groups find it very difficult to anticipate major trends 20 to 30 years into the future. The strategic planning itself may spur decisions that alter university policy and behavior of its faculty, staff, and students.

Characteristics of successful strategic planning include commitment to planning by top management, continual scanning of the external environment, regular monitoring of internal data, inclusion of all stakeholders in the planning process, and adoption of implementation tactics.

This chapter outlines key aspects of a process-oriented strategic planning model.[1] It is a model that has evolved through working with five topical strategic planning groups during the time I have served as Vice President for Academic Affairs of the University of Wisconsin System. Colleges of business and engineering have made major changes and developed major budgetary initiatives following this model; guidelines for colleges of education will be implemented later this year. Planning groups for academic libraries[2] and student affairs are using the process to tackle issues in these important educational services.

TOPICAL STRATEGIC PLANNING: THE CONCEPT

Strategic planning as practiced in the University of Wisconsin System asks representatives of various campuses of the system, from state agencies, the public sector, and citizens groups to focus on a topical area. The group studies current practices and trends in a field of knowledge and develops guidelines to chart change from the system to the institutional and individual levels. Programs and practices that anticipate the future often become models for new guidelines. Barriers to maintaining or achieving quality in programs of research, instruction, and public service are identified. Recommendations developed and agreed to by the planning group are reviewed by faculty, staff, and administrators at the institutional level and the academic vice president and the president at the system level, before presentation to the governing board for approval.

As we have implemented it, strategic planning involves members of the university community, but reaches beyond it to the citizenry. One major unanticipated benefit has been the increased understanding on the part of the citizen planners of the issues facing the university and the procedures already in place to tackle them. Another is the increased awareness on the part of members of the university community of the citizenry's perception of the university. Symbolic and substantive changes generated by the process resound well for higher education in the state.

The model relies on bringing knowledgeable and creative people together around a charge that outlines a general process and asks them to produce a set of guidelines to help institutions and the system plan with a 20 to 30 year horizon. The particular procedures used to generate the information and the recommendations rest with the members of the planning group. To provide an element of commonality for the planning group, I invite several experts from outside the University of Wisconsin System to present papers outlining major trends and issues at a one to two day conference to which additional people from throughout the state are invited.

The other key players in process-oriented strategic planning are the senior academic administrators of the institutions of the system and the chair of the faculty senate for each institution. The vice chancellors of the system campuses identify topical areas, review the prospectus, nominate candidates for the planning committee, suggest outside consultants, participate in the conference, receive periodic updates on the work of the planning group, and review its draft report. Like the vice chancellors, the representatives of the faculty receive periodic updates and reports as the process moves along. In sum, topical strategic planning as we practice it is an open process that permits and encourages involvement beyond the planning group.

Topics for strategic planning efforts are developed on a consensual model. Once a potential topic is identified, it is discussed with administrators, faculty, and staff responsible for it at the institutional level. These discussions often help to crystalize the issues and frame the questions for the charge to the planning group.

The agriculture and natural resource experience illustrates the concept of topical strategic planning as we have developed it in Wisconsin. The System had not engaged in long range planning in agriculture and natural resources since 1974. In the intervening fifteen years, much had happened in these two broad areas. New research and policy themes had emerged and attracted public attention. Undergraduate and graduate enrollment in colleges of agriculture and natural resources had been volatile with some degree programs remaining stable while many declined. Although the student population and the faculty have become more diverse, they did not reflect the heterogeneity of the general population. Farm families and high school agricultural programs, traditional sources of students and faculty, had declined in number and in enrollment, respectively.

Clearly strategic planning was timely; academic leaders at the institutional level agreed. A prospectus was developed, reviewed, and approved. I named an eighteen member planning group from a list of nominees submitted by chief academic officers of the institutions of the system. A chair was appointed and an academic planner from my office designated to serve as staff for the group over the approximately eighteen months estimated for the planning group to produce a draft report.

TOPICAL STRATEGIC PLANNING:
STEPS IN THE PROCESS

Topical strategic planning in this model is a process that moves from issue identification through resource inventory to establishment of guidelines for programmatic direction and organizational change to more effectively deliver the teaching, research, and outreach functions under future conditions.

To achieve these objectives, the first task of the planning group is to define issues and distill them into a synthesis of future trends in the topical area. This book is testimony to the enormity of this task. The papers that appear in this volume were commissioned as an initial step. They were presented at a conference to which all fifteen institutions of the system named delegates who participated in the discussion groups and interacted with the speakers. In addition, the planning group divided itself into subgroups, and invited experts from the state to identify trends and share data. The collective output of these two activities is the synthesis of trends in agriculture and natural resources and their intersection that follows in Chapter 2.

A second major task is to consider the implications of these trends for instruction, research, and outreach. Instruction illustrates the magnitude of this task. The rapidity of the creation of knowledge and the studies that challenge how well we have taught critical thinking and problem solving skills necessitates some attention to the type of preparation needed for future graduates to successfully perform as citizens and professionals in the twenty-first century. In some areas, agriculture and natural resources for example, new ways to expand the supply of students receive attention. Overall decline in enrollments, a precipitous drop in certain fields, a potential future shortage of faculty, the ability to continue to attract talented students, and the continued underrepresentation of women and minorities in almost all fields are a few of the major questions that planning groups have no choice but to address.

A set of planning guidelines is the third, and final, task of the planning group. Institutional micro-management is not the objective. Rather than a detailed plan, the document should establish a set of goals, and strategies for their attainment, that are embedded in a framework that identifies opportunities and trends, and acknowledges institutional strengths and internal and external constraints. Typically, a working draft of the guidelines is sent to a panel of in-state experts for comment.

The planning group for agriculture and natural resources invited comment from the conference participants.

The planning group's report is submitted to the academic vice president of the UW System. The system president and academic vice president involve the institutions and the public in their review process. The formal review process at the institutional level is handled by the chief academic officer. Representatives from outside the university provide recommendations for external review. The recommendations presented by the president to the systemwide governing board for action reflect those of the planning group and comments from faculty, staff, administrators, and the citizenry.

Topical strategic planning is not a part of the budgetary process, but guidelines developed by the planning group and accepted by the president and the governing board inform budget priorities. The accumulation of individual decisions to allocate resources according to the priorities established in the guidelines shifts resources at institutional, college, program, and departmental levels. The planning process may also set the stage for successful requests for new monies.

CONCLUSION

The ability of a system of higher education to respond to a dynamic world requires creative, flexible, and aggressive programming. Topical strategic planning can produce the vision and the actions necessary to meet the challenge of excellence and service.

Topical strategic planning depends on those of us at the system level to create conditions that will enable a knowledgeable group of people to envision the future and develop broad guidelines suggesting how institutions respond. The process and the product shift the vision of the landscape.

By giving individuals who are interested in the issues a voice in the process, commitment to a new perspective emerges. Programmatic, curricular, personnel, and budget decisions by faculty committees and administrators take into account the new vision. Individual members of the faculty and staff carry that vision to the level of professional choices—research topics, linking research and public service, involving undergraduates in research projects. Strategic planning is connected to other planning efforts.

The process is the genesis of change.

NOTES

1. General difference between this approach and that of several other universities and university systems as well as the connection between strategic

planning and other types of academic planning is discussed by Robertson in Strategic Planning, a paper presented to the Board of Regents in April 1990.

2. The papers from the Strategic Planning Process for Libraries have been published as "The Future of the Academic Library" in the University of Illinois at Urbana-Champaign Graduate School of Library and Information Science Occasional Papers series as numbers 188 and 189.

2

Agriculture, Natural Resources, and the Environment

WAVA G. HANEY
DONALD R. FIELD

THE GLOBAL CONDITION

The environment is a complex web of interconnecting ecosystems and human activity. Economic activity is of special interest within this web. Recently, it has become clear that local economic activity can affect ecosystems both regionally and globally. Similarly, a local ecosystem is affected by activities happening in other parts of the region or the world. In the words of the World Commission on Environment and Development (Brundtland and others, 1987), "We are now forced to accustom ourselves to an accelerating ecological interdependence among nations. Ecology and economy are becoming ever more interwoven—locally, regionally, nationally, and globally—into a seamless net of cause and effect." Therefore as we close the century, it is clear that natural resource concerns are important issues that establish the context for and will substantially influence Wisconsin agriculture from now to the year 2020.

The state of the global ecosystem is central to maintaining production systems. People use natural resources and consume agricultural products. More people require more resources, thus population growth is one basic driving force in global environmental issues. Since their population is growing at a faster rate, Third World countries will comprise an increasing proportion of the world population. With more demand placed on diminishing resources and limited financial means to protect the environment and respond to pollution problems, decisions made in less developed countries play an important part in the state of the global ecosystem. However, industrial and industrializing countries, through their high consumption, may have the greater consequence for the global resource base. Unless developed countries use their knowledge and wealth to balance consumption with resource management, the global ecosystem will remain in jeopardy.

Changes in the chemistry of the atmosphere affect vegetation, soil, and water. The implications of atmospheric changes are unknown, though the effects of some of these changes are thought to be potentially great. For example, without a significant reduction in nitrogen oxide and sulfur dioxide emissions, acid deposition could impact agriculture, forests, fisheries, lakes, rivers, and human health. Global climate change could alter local weather patterns and sea levels, and thus produce shifts in agriculture production among different domestic and international regions.

The biosphere also warns of serious problems. Species reduction could limit the source of genetic material for improving strains of common food crops, the source for development of "new" crop strains, and the source of active ingredients for about 40% of prescription drugs on the market today.

Many global environmental concerns—deforestation, desertification, and loss of productive land and critical ecosystems—are ultimately land use issues. Land use is central to many global environmental issues because decisions are often virtually irreversible and their repercussions felt both locally and globally. Reconciliation of competing demands for the preservation of critical ecosystems and their development for land uses supporting economic activity is a major concern for scientific study and public debate. Land stewardship and identification of uses that do not exhaust the productive capacity of different types of land are related concerns. Natural beauty and the attractiveness of the landscape also help to define the quality of life for Wisconsin citizens, seasonal residents and tourists.

The energy issue today, and into the future, is the transition to the next generation of energy supplies beyond fossil fuels. Future sources of energy have implications for natural resource management affecting renewable energy sources. This has significant implications for agriculture since energy is a major agricultural input and some agricultural products and by-products are potential sources of energy. Changes in energy sources affect agricultural practices, agrichemical production and distribution, and food processing and distribution.

In the past twenty years, a growing awareness of the adverse effects of many by-products of human activity has resulted in a call to control the discharge of conventional pollutants into the air and water, and the disposal of solid wastes on the land. Toxic and hazardous waste management adds new dimensions to air and water management and emphasizes the critical nature of residuals management, including recycling and source reduction. The technology to store, manage, and dispose of these wastes is costly and sophisticated. Detection and remediation are beyond the capacity of most localities. Given the amount of resources necessary to properly manage toxic and hazardous materials, waste management and disposal will continue to be major issues at all levels from local to global.

Side effects of food production is another environmental issue of the 1990s and beyond. Major public health and policy issues include contamination of groundwater by nutrients and pesticides, pollution of surface water with sediment and chemicals, and hazards that pesticides and other agricultural chemicals may pose to the health of farmers, consumers and wildlife.

Agriculture must play a central role in the environmental discussions. U.S. agriculture is envied throughout the world for its ability to improve productive capacity and provide a quality and wholesome food supply for an ever expanding world economy. While agriculture has an obligation to provide food and fiber for the growing world population, that obligation must be met in a manner that does not deplete or destroy the natural resource base on which it rests.

Like forestry, fisheries, mining, and tourism, agriculture can no longer pursue a resource policy in isolation. Vital systemic relationships exist between forestry and agriculture, agriculture and fisheries, primary resource production processes and resource protection.

In sum, healthy lands, water, forests, wildlife, fisheries and communities are essential to agricultural production and resource conservation. These resources form the framework around which rural life has evolved. The people who farm the land, harvest the forests, and enjoy the bounty of nature are actors in a dynamic, global ecosystem, subject to the social and biological forces which regulate these systems.

The words that follow attempt to capture the uncertainty, challenge, and excitement of a bright new era for agriculture and natural resource programs.

AGRICULTURE IN A GLOBAL CONTEXT

Agriculture and food production became an integral part of a dynamic world economy during the twentieth century. Worldwide application of agricultural research, science and technology brought self sufficiency, adequate diets and improved nutrition to many nations. Unfortunately, severe food production and distribution problems still plague developed and underdeveloped countries on several continents. Current food shortages and distribution problems in Africa, Eastern Europe, Latin America and the Soviet Union, amplify the critical need for expanded agricultural research, education and development.

The development of global commodity markets has had a significant impact on the market structure of U.S. Agriculture. The internationalization of the U.S. food system has occurred rapidly in the food manufacturing and food retailing industries. Most of the large U.S. food marketing companies are multi-national firms closely linked to key trading centers in Europe, Asia and South America. The industrialization of the U.S. food system requires favorable trade negotiations and foreign policy to provide open access to international markets. The viability and profitability of the American farmer is often directly related to agreements and concessions in the General Agreement on Tariff and Trade (GATT) negotiations, world monetary policy and the stability of foreign governments.

Food production and agricultural trade will remain a significant part of the world economy in the twenty-first century. Population expansion, improved nutrition, growing affluence, changing demand and the domestic food policies of many countries will continue to produce rapid changes in prices and demand for U.S. food production.

Technological advances in biotechnology, genetics, integrated pest management and food technology will help ensure a viable food production system. Scientists will be challenged to explore new frontiers of knowledge and develop new technologies to provide the productive capacity needed by a global society.

The dynamic changes in agriculture are not only technological and economic but social and cultural as well. In 1989, academic and industrial leaders of the National Research Council urged that the definition of agriculture be broadened beyond the system that grows, processes and distributes food and fiber to "also encompass the related natural resource industries, public policy issues, social systems and physical and biological environments." Structural changes in the global economy will continue to reshape all sectors of the U.S. economy including non-metropolitan communities and the agriculture and natural resources industries on which they depend. This dependency of rural areas on goods-producing industries like agriculture and forestry will increase the ties between many non-metropolitan counties and the global market place.

The growing interdependence of the world economic community has also brought new challenges and opportunities to agricultural colleges and universities. Highly trained graduates with knowledge and understanding of international markets and with multi- and bi-lingual skills will be in strong demand. International research will require students capable of interpreting cultural differences, socioeconomic policy and geographic data. Scientific breakthroughs will accelerate as the research and educational communities focus on common objectives and basic needs.

American agriculture will continue to play a significant role in the world market economy in the next thirty years. U.S. agribusiness is the world's largest commercial industry with over a trillion dollars in assets and 20% of the national work force. The total number of harvested acres has remained constant over recent decades, but the actual number of farms has declined sharply. About 15 to 20% of the commercial farms produce over 80% of the national food supply, while three-fourths of the farm households now have supplemental off-farm income.

WISCONSIN'S AGRICULTURE TODAY
AND TOMORROW

National data often mask regional variations. In the Wisconsin-Minnesota dairy industry a relatively small proportion of milk production comes from very large commercial farms. The farm population here is generally more dependent on farm income than in many parts of the country, and fewer than 30% of the farmers hold parttime jobs. In Wisconsin, there has been a growing trend to multi-family farms with greater acreage and larger herds.

Agriculture and food processing, distribution and retailing provide a significant source of employment and income in nearly every community in the state. Commercial farming is a major industry, producing over $5 billion annually in cash farm receipts and providing the raw products for Wisconsin's extensive food processing industry. Nearly 85% of the state's agricultural products are consumed outside the state, with significant amounts of commodities and services marketed in international trade. Dairying is the state's major agricultural industry, but Wisconsin is also a leading provider of many vegetable crops that are processed in the state and marketed domestically and for export. Both producers and processors are considering an array of alternative vegetable crops and fruits to meet changing consumer demands.

The critical challenge for Wisconsin agriculture in the next few decades is to stay competitive in national and world markets. Meeting that challenge will necessitate balancing efficiency and profitability with environmental integrity. Global environmental and economic trends underscore the rapid rate of change and the scope and complexity of factors producing change. Six major factors will shape the trends. They are:

1. the continued development of global capital and product markets;
2. the content and degree of coordination of economic, social, and food policy worldwide;

3. the emergence of mega markets, like the European Economic
   Community and the Pacific Rim, and the breakdown of market
   barriers;
4. population growth and population shifts;
5. the impact of global climate change, droughts and floods;
6. and the growth of biotechnology and the explosion of technical
   and scientific knowledge generally.

Within this context, some of the specific challenges and opportunities
facing the agricultural community in the next few decades are: developing
new national and international markets for our products and new products
for international markets; strengthening and increasing cooperative marketing; upgrading food processing plants by adapting new food engineering technology; increasing agricultural input options and input
systems that protect and enhance soil, water, and other fragile resources;
helping producers and processors develop superior business management
and technical skills; assisting the food industry to develop ways to
compensate for Wisconsin's location and climate; developing and transferring information and new technology related to production, processing,
and marketing; expanding technology transfer to emerging service sector
occupations ranging from crop consultants, tree disease specialists, and
licensed chemical operators to food servers in the fast food industry;
and developing policy options that will encourage profitable production
while maintaining a safe environment and meeting demands for high
quality, safe, and nutritious food.

FOOD SAFETY AND HUMAN NUTRITION

Although the United States food supply is among the safest in the
world, public concern about food hazards has risen dramatically in
recent years. New technologies may increase production efficiency and
reduce producer risk, but also give rise to consumer concerns about
harmful residues in food and changes in food properties. New processing
and marketing technologies can enhance food quality, shelf life and
consumer convenience, but also may increase the possibility of food
safety problems. Biotechnologies aimed at improving food production
and processing are of particular public concern. At the same time,
sophisticated instruments have greatly increased the ability to detect
pesticides, chemical residues, toxins and harmful microbes in food.

Added to this rapidly changing food safety scene are numerous activist
groups and organizations that have found new ways to bring food
hazards to public attention. Increased activism has led to demands for
new public policies to protect consumers from such perceived hazards.

Neither policy demands nor perceived risk can be ignored. While new technologies can improve detection of risks to public health, improvements in the communication of risk information to decision makers and the public can inform public policy development. A new outreach role is being formed around human nutrition directed at new clientele, ranging from those interested in physical fitness to those who serve prepared foods, as the public relies increasingly on fast foods from restaurants, delicatessens, convenience stores and other vendors.

The twenty-first century will see no let up in consumer demand for improvements in food quality and convenience. Changes in lifestyle, family structure, and the ethnic and cultural composition of the U.S. population will continue to have an impact on dietary needs and food product demand. Dual-income and single person households will increase the demand for good tasting, nutritious food that is easy to prepare or readily available.

On the nutrition side, American consumers have more information on healthy eating habits available for their use than ever before. Yet, these consumers remain susceptible to dietary fads, and surveys suggest that the public lacks an understanding of the role of food in human health. Additional effort will need to be devoted to convey nutrition information to the public, and attention must be focused on the barriers which inhibit individuals from acting on that knowledge. Research is underway that will increase understanding of the relationship of diet to general health, longevity and chronic disease. Specific diets are being developed to meet health needs of particular age, gender and disease groups. The need to provide adequate nutrition to the growing stratum of economically disadvantaged, especially children, will be a major challenge. Fitness and prevention of health problems will encourage continued attention on the nutritional value of foods (e.g., calorie content, cholesterol, types of lipids, sugar, salt).

Overall, food quality is likely to continue to improve. Biotechnology offers the prospect of genetically engineered plants and animals with increased disease resistance, faster growth, greater productivity, and the possibility of altering other plant traits that can improve the nutritive value of foods. Genetically engineered enzymes for more effective food processing, synthetic and reshaped foods (e.g., leaner pigs, low cholesterol eggs), new and/or improved food ingredients, and gene probes to detect food pathogens such as Salmonella and Listeria are other possible contributions of biotechnology. New analytical methodology can enhance the ability to detect small concentrations of specific substances in food and the environment, broaden the range of substances that can be detected, and identify new types of food spoilage. As it will be impossible to create a totally risk-free environment, the technologies for detection must be coupled with increased attention to risk assessment.

Ability to detect new substances can lead to changes in food processing, handling, packaging and preparation. New food packaging and packaging materials are likely to be developed that will permit food to be microwaved without the migration of materials from package to food. With such technologies, safe consumer handling and preparation of foods will continue as major food issues of the future.

Food safety has increasingly important economic consequences. With greater consumption of food outside the home and with more convenience foods purchased in prepared form for in-home eating, consumers are able to change their demand for food much faster than in earlier decades. Thus, any actual or perceived problems in food safety, combined with current media capabilities, can have a large and immediate impact on the demand for a food commodity with severe economic consequences on a portion of agricultural producers and rural communities.

NATURAL RESOURCES AND THE LAND ETHIC

There is increasing awareness that natural resources undergird our most basic production practices in agriculture, forestry, fisheries, wildlife and water resource management, and energy and mineral development from the local to the global level. In recent decades, recreation, tourism, and aesthetic vistas have been included as natural resources requiring management. Airsheds have been added to watersheds as contemporary natural resources that may well influence global climate change, agriculture, forestry and mineral production practices and human settlement patterns in the future.

The management of these natural resources for human welfare has become a driving force in public policy (e.g., Clean Air Act, Clean Water Act, Food Security Act, concerns about global species diversity). Environmental stewardship and resource conservation are central themes.

Both consumptive and non-consumptive natural resource uses are of immense economic importance to Wisconsin. The forest products industry, led by paper, employs about 77,000 workers who convert $67 million worth of trees and recycled fiber into shipments worth over $10 billion annually. The mining industry sells $232 million worth of raw materials annually to be converted into numerous final products.

The tourism industry, fragmented into a multitude of activities and services that are provided by numerous small businesses, includes about 100,000 jobs and has $4.5 billion of economic impact per year. Sport fishing involves 1.8 million anglers who spend $712 million and generate another $1.2 billion in annual business activity. Over 700,000 hunters spend $311 million while 3 million non-consumptive users of wildlife spend $338 million per year. Separate economic impact figures are not available for other non-consumptive uses such as auto touring, bicycling, boating, camping and skiing, but marinas on the Great Lakes, for example, generate $84 million per year.

The magnitude and diversity of natural resource industries have further escalated resource policy questions and refined the resource policy process. While new natural resources may continue to be discovered, the diverse and conflicting demands for existing resources will dominate resource management over the next 30 years. The major trends and issues noted here reflect the debate about the "choice among alternatives" that requires a delicate balancing of economic development and natural resource conservation considerations.

The policy struggle with regulations regarding export of raw materials and value-added processing will escalate as production sectors become distributed throughout various world economies. Forest resource production and marketing operations will concentrate on competitive advantage and considerations for alternative uses of resources in the context of water quality, tourism and wildlife and fish habitat. Ecosystem management principles for both the private and public land manager will become standard operating procedures. The unit of analysis will shift from field, farm and forest to a macro landscape system.

Similarly, non-consumptive use of forest and other natural resources will increase. Both recreational and urban forests will grow in number and importance for public land management agencies. Aesthetics, catch and release fisheries and non-game wildlife programs will dominate recreational thinking, activities and management.

With greater diversity in expectations from a limited resource base and from participants in tourism and recreation activities, conflicts over resource management will accelerate. Debates on the purpose of public forests, surface water use and the recent Indian treaty rights issues are examples of current resource use conflicts. Disagreements will continue between proponents of individual and collective rights and advocates for the public good. Specialized and exclusive recreational uses; motorized recreation/high technology equipment; noise and litter; lack of stewardship by landowners; short-term perspective of investors; and regional versus global benefits are some of the most likely topics of conflict.

The degree of government regulation of private property and business in the areas of land use planning/zoning; billboard control; waste management and treatment and waste load allocations of pollutants; impact of regulations on small businesses; and biotechnology in forestry and wildlife will be debated seriously.

Institutional arrangements will be developed to internalize the costs and benefits from the use of natural resources such as wildlife damage to agriculture and conversely agricultural damage to wildlife. Integration of the aesthetic and ecotourism benefits from pastoral landscapes and wild areas with other resource uses will be supported by agency regulations and programs.

Finally, questions of philosophy and ethics will be dealt with more directly by the universities, management agencies, private businesses and the citizenry. Focused clientele relationships will become more holistic and long-term. A philosophy of responsibility for future human generations to non-human members of the biosphere and the air, water and soil—the land ethic—will be more clearly articulated.

RURAL PEOPLE, COMMUNITIES, AND QUALITY OF LIFE

Rural America is diverse. Its communities show great variation in sources of employment and income. Farming, forestry and mining are the heart of many rural community economies, but they are generally declining in importance. Employment in manufacturing, construction, tourism and other services, and income from transfer payments, interest, dividends and rent provide the major sources of personal income in many non-metropolitan counties. Rural America, like our modern day cities, has pockets of prosperity and poverty.

Rural Wisconsin residents make up 30% of the state population. They derive their income from manufacturing, tourism, retail and service businesses, agriculture and forestry. Rural Wisconsin, viewed across communities, is extremely diverse in terms of culture and income sources. Rural communities have been hit hard by structural changes in the world economy. Greater dependency among economies links non-metropolitan America to the health of agricultural exports and factors such as international capital markets, food policies in other countries and the manufacturing market. The diffusion of technology allows nations with less skilled and less costly labor to undertake production of standardized goods in a more cost efficient manner. This puts the United States at a cost disadvantage and many of these manufacturing jobs are moving to other shores. Communications technology and the containerization revolution in transportation will continue to allow the production process to be fragmented and small nations to specialize in components that are then brought together from different parts of the world to one point where they are assembled.

Hence, many of our rural communities are at a disadvantage. The fortunes of rural areas, that showed significant improvement in the 1970s, took a turn for the worse in the 1980s. While growing slightly faster than urban areas, population growth in rural Wisconsin hovered at 2.9%, but has slowed. Employment growth has been less than one third that of urban areas. The unemployment rate is higher in rural areas—a reversal of conditions during the 1970s. In 1986, the average urban unemployment rate stood at 5.8%, while the rural unemployment rate was 7.2%. That same year the average Wisconsin rural resident received $2,500 less than the average U.S. per capita income, while the average Wisconsin urban per capita income exceeded the national average by $300 (Wisconsin Department of Development, 1988).

Educational attainment is lower, more housing is substandard, fiscal capacity of local governments is lower, and the poverty rate is higher in non-metropolitan areas than in metropolitan areas. The gap between the standard of living of rural and urban residents has grown during the recovery and expansion phases of the current business cycle.

Higher education can make a difference in rural Wisconsin. It can help diversify and expand the economic base; assist citizens to optimize output from current resources; produce and make available knowledge to improve rural conditions; advise in changing the existing organizational framework (infrastructure and delivery systems) to improve performance of basic institutions (e.g., local government, school systems); instill a desire/urge to change; develop a cultural environment that allows for discussion and debate; and tap the significant unused potential for leadership among the highly trained women in rural areas.

SUMMARY

Without a doubt, agriculture and natural resources have entered a new era. Their context shifts rapidly and sometimes in unpredictable ways. Human capital for agriculture and natural resource sciences and programs is also in transition. The next three decades will find a workplace and work force that differ significantly from the past.

The challenge to Wisconsin's educational system is clear. The university of tomorrow will find more professional women and minorities occupying faculty roles and helping to redefine academic affairs. According to one estimate, academic recruitment between 1990–2004 will require 335,000 new faculty members representing the social, biological and physical sciences, arts, and humanities (Creswell and others, 1990). This will lead to a new university as we enter the twenty-first century. Similarly, agriculture and natural resource production, marketing and distribution occupations in the private sector will rely on more women and minorities. The tensions between the shortage of work force participants and skilled workers with a mastery of emerging technologies and the introduction of increased numbers of women and minorities in the workplace will dominate management thinking throughout the decade.

II

Agriculture and Natural Resources Futures

OVERVIEW

The stage is set. An outline for current issues and future trends is defined. To provide a context for Wisconsin's interpretation of contemporary and future events affecting agriculture and natural resources, senior scholars from government and the academic community gathered in Madison, Wisconsin in November 1989, to provide advice and counsel. Their collective analysis in the chapters that follow provide an overview of national and global trends and issues facing agriculture and natural resources and help to define goals and strategies for meeting the needs of the present and adjusting to the demands of the future.

Topics progress from domestic to global. Charles Benbrook outlines the general state of agriculture and critical issues facing the nation while tracing the implications for Wisconsin. The breadth of his views provide a national perspective, and the topics covered clearly convey the complexity of the problem and the multidisciplinary and multi-institutional partnerships that will be required to address them. While outlining the opportunities and challenges for the Land Grant University System, he raises questions regarding the capacity of its educational system to address such issues. He suggests that a range of federal and state agencies, citizen groups, and educational leaders need to work together to evaluate and build effective public policy.

While Benbrook carefully identifies domestic issues, Ann Tutwiler focuses on global ones. Drawing upon her background in international relations, she argues for a shift in assessment by the agricultural community from fixed exchange rates and protectionist policies toward more integrated world markets and expanded trade practices. Her macroeconomic analysis shows the importance of increased trade between developing and developed countries to the future of U.S. agriculture.

The evolving global economic environment requires different types of training for future agriculture and natural resource leaders. This is the message accentuated by Duane Acker. Focusing on production and marketing issues, Acker argues that U.S. agriculture must pay more attention to the interests of consumers in domestic and international markets. Food habits are changing rapidly. If U.S. agriculture is to be competitive in world markets, consumer taste preferences and cultural variations in food service and consumption patterns must be incorporated into product development and distribution. Both Tutwiler and Acker call for U.S. agriculture to increase its assistance for food production in developing countries. Their rationale hinges on the argument that a strong food production sector increases trade capacity.

John Gordon applies the agriculture themes of Benbrook, Tutwiler and Acker to forestry. Like Benbrook, Gordon sees a strong stewardship and environmental theme emerging within the forestry profession. He suggests that fiber production capacity must increase to accommodate world needs, but within the context of a more rationale system for evaluating competitive wood product advantage that allows for profitability while protecting natural resources. Sustainable forestry would integrate fiber production, multiple use, and resource protection. This transformation would be based on need paradigms operating to help construct future forest policy.

Like Tutwiler and Acker, Gordon sees an international future for U.S. forestry with accompanying challenges to the profession. He suggests that a new forestry education agenda must be charted that broadens the scientific skills and the knowledge base of the profession. He also recommends alternative educational options for the professional forestry program. Gordon provides a proactive-prescriptive approach for tomorrow's forestry profession and industry.

Mark Lapping elaborates on Gordon's emphasis on landscape planning. He begins by defining the important components of this planning process: farm land preservation, land ownership, land use patterns, size and scale of farm operations and labor force characteristics. These must be incorporated into regional systems to provide operational criteria for planning and managing the landscape. Lapping's concern for maintaining the integrity of land through quality management practices and rural infrastructure development is carried throughout his analysis of current and future agriculture and natural resource issues.

James Zuiches shifts the focus to the human component of the rural landscape by examining the institutional infrastructure. While applauding the increased production capacity of farm operations, he points out that the ensuing transformation of agriculture has left many rural communities and regions on the verge of collapse. Zuiches argues for greater attention in land grant colleges to social science data about rural communities and rural people. His call for integrating socio-economic information with agriculture and natural resource production data could enhance policy alternatives for sustaining the human community in rural regions.

Mary Heltsley echoes Zuiches's concern for the rural population and their communities. She provides a complementary perspective on the quality of rural life, drawing attention to rural families, rural youth, and the rural elderly. Heltsley catalogs many of the social problems facing rural residents including the lack of employment opportunities for women and the resulting loss of potential leadership capacity for many rural communities. Her comments on youth at risk and the problems of the elderly draw further attention to the array of needs of the people who reside in rural America.

Russell Peterson provides a capstone to the various perspectives on issues and trends in agriculture and natural resources. His is a global perspective that cites numerous examples of the interconnectness of the physical, economic, and social realms of the biosphere. Illustrations of the limits of our natural resource base preface his challenges for the future. In calling for renewed attention to world resource management, he accentuates recommendations on soft energy systems, new technology and conservation practices for a global citizenry. He translates his view of the biosphere into educational requirements for a new generation of scientists and policy makers. He culminates his remarks with a plea for interdisciplinary thinking and training and with specific recommendations for academic programs.

Together these national leaders have provided individual wisdom and collective thought on the future of agricultural and natural resource education. Their respective challenges, as outlined in the accompanying chapters, underlie the recommendations made in the final chapter.

3

Tapping the Full Potential of Wisconsin's Human and Natural Resources

CHARLES M. BENBROOK

I begin with the premise that the mission of the University of Wisconsin System is to improve the quality of life for the citizens of the state by creating and deploying knowledge and fostering reason in efforts to make the world a better, safer place. The tools of the trade for this great university are education, research, and involvement in both public and private decision-making processes that impact the general welfare of citizens in the state. Involvement can occur in many ways, including institutional and individual faculty and student activities.

My mission is to assess emerging challenges for the university system arising from and confronting some of the state's basic natural resource-based industries—farming, food processing, forestry, and recreation. Many people wonder why agricultural and food policy issues, particularly those dealing with the environment and food safety, have become so much more prominent in recent years, and at times so divisive. The reason can be traced, I believe, to growing recognition among all people that agricultural, food, and forestry industries profoundly impact our quality of life through the food we eat, the water we drink, our health, and the cultural, natural, and economic environment we live within.

I turn first to challenges the university system will have to face in the 1990s. My focus will be what I believe to be the most pressing, difficult issues that both students and citizens in the state will expect this university to help solve through education, research, involvement, and outreach. My remarks are organized into four parts: natural resource stewardship, food safety and quality, social expectations and economic performance, and meeting global food needs and market opportunities. In light of these challenges, systemwide and institutional strengths and weaknesses are surveyed. Last, I suggest some opportunities, and point to some imperatives for change.

EMERGING CHALLENGES

Longstanding challenges already lie at the core of many university programs, and will surely persist throughout the next century. New ones will emerge, and old ones will take on new dimensions. Science will both change the public's perception of problems, and create new opportunities to pursue novel solutions. The role and activities of public institutions will evolve, guided by the judgments and perceptions of citizens regarding what steps or resources are needed to best serve the public good.

One difficult question may arise with increasing frequency. Can society's most pressing problems be traced to a lack of knowledge or a collective inability to fully or effectively use existing knowledge? A crisp answer to this question is likely to prove elusive. Moreover, the answer is likely to vary greatly in different areas of social endeavor, as will the appropriate steps to adjust the balance of research and education activities in accord with emerging needs. Nonetheless, it is a key question to keep in mind in assessing the future role of the university.

My own sense is that we collectively approach the generation of new knowledge more aggressively than fostering full and proper use of existing knowledge. In agriculture, we are compelled to invest more and more resources trying to solve many of agriculture's most troubling biological production problems, when the logical solution might be to shift production patterns and technologies to simply eliminate the most intractable problems. When we fail to recognize and pursue such solutions, we are failing to make full use of existing knowledge.

Natural Resource Stewardship

Natural resource issues are, in my judgment, among the most clearly defined challenges confronting the university. Even so, they are likely to be among those the university system is least able to effectively confront in the years ahead. As a result, the role of the state's natural resources and agriculture departments, the federal Environmental Protection Agency and U.S. Department of Agriculture, and other public and private institutions can be expected to grow, and will clearly impact the appropriate role of the university. If the public believes these other institutions are effectively responding to current needs, the claim of these public institutions to public funds may grow relative to that of the university. I might add that much more is at stake than the role of Extension.

Water Quality. Fresh water resources are among Wisconsin's greatest natural treasures. The state is blessed with lakes, including hundreds of miles of shoreline along the Great Lakes, rivers, streams, and wetlands, as well as accessible, high-quality groundwater resources. Access to high quality water is one of the key pillars of the state's economy. While there is tremendous potential to more fully utilize this natural endowment, there are also troubling indications that future opportunity will be eroded by the current generation's inability to more effectively protect water quality. Indeed, some water quality problems around the state may emerge as so serious in the public's eye that tough choices may be difficult to avoid in the next decade.

The tangible benefits to the state from wise use of water resources, coupled with determined efforts to prevent pollution will flow from many sources. These will include:

- lower costs to meet urban and rural demands for water;
- opportunity to expand industrial activity dependent on affordable access to high quality water (including certain manufacturing industries, breweries, food processing, livestock agriculture); and
- improved commercial fisheries and recreational opportunities.

The university system is already deeply involved with helping the state meet the challenges inherent in managing this vital natural resource. Across the nation, though, we continue to lose ground in some key areas. Pressures are growing at all levels of government to pass more all encompassing water quality laws, and to more aggressively accelerate progress in meeting already codified water quality protection goals. Knowledge and reason clearly are needed if society is to channel these pressures into constructive public solutions that do not (pardon the pun) throw the baby out with the bath water.

Water quality is in the process of emerging as agriculture's major environmental challenge. Two problems are uppermost on everyone's mind—the leaching of nitrates into groundwater and other sources of drinking water and pesticide contamination.

Nitrogen contamination poses by far the most ubiquitous near-term challenge. Recent evidence strongly suggests that high risk regions are characterized by high concentrations of livestock, extensive production of forage legumes, and intensive row cropping systems—regions like much of Wisconsin.

In areas where farmers have to pay cash for most of the nitrogen needed to support crop growth, economic realities tend to limit the frequency of clearly excessive rates of fertilization. In regions where farmers have large quantities of manure to dispose of on land routinely planted to legumes in rotation, it is not uncommon for grossly excessive amounts of nitrogen to be applied. Researchers at Madison are well aware of areas in the state where some farmers apply 75 to 125 pounds of nitrogen fertilizer onto land that needs *no* or only modest levels of additional nitrogen. That's why nitrate levels in groundwater in parts of the state are approaching or already exceed 10 ppm—the EPA standard for public health protection.

The management of manure must become a much higher state-wide priority. Manure is both an untapped resource to farmers and a costly pollutant when nutrients wash into surface waters or leach to groundwater. Fortunately, the economic benefits to farmers from improved nutrient management often will largely or fully finance the investments and costs associated with adherence to Best Management Practices (BMPs) for manure management. Excellent research undertaken by the university has demonstrated these opportunities, yet many farmers have still not made the needed changes in their manure management programs. Does anyone know why? Has a systematic effort been made to find the answer to this question?

In protecting water quality from nitrate contamination, a first essential step is fostering better management of manure. A second step is to encourage farmers to more carefully monitor and calibrate soil nutrient levels and cycles, and to give them the tools to do so. Then, on the basis of this improved information, farmers must be persuaded, bribed, or penalized into cutting back on fertilization rates when and as appropriate in light of projected crop needs.

No one knows whether widespread adherence to BMPs for nutrient management will fully solve, largely solve, or just partially solve nitrate contamination problems. Some people think it makes sense to find an answer to this rather basic question as soon as possible, and hopefully prior to the imposition of costly and divisive new regulatory restrictions or penalties on fertilizer, manures, or other sources of nutrients.

While important challenges face the agricultural sector in managing agricultural wastes, there is also great potential to utilize and recycle urban wastes as a rural resource. The costs of disposing garbage of all sorts are escalating across the nation, and new technologies and recycling strategies are emerging which will expand the range of opportunities for farmers to utilize urban waste as a source of nutrients and organic matter.

Despite the often more emotional contemporary concerns arising from pesticide contamination of water, these problems are likely to be more readily resolved than nitrate problems. Farmers in Wisconsin have many viable pest control options, some of which involve little or no reliance on pesticides. But they have yet to learn how to grow corn, small grains, and other crops without nitrogen.

Moreover, not all pesticides have the potential to leach into ground-water, and only a handful appear to pose serious risks as a result; and unique hydrological features tend to characterize regions prone to ground-water contamination. As a result, the solutions to pesticide-related water quality problems are likely to be more self-evident, and need not pose serious costs on farmers or society, unless of course we fail or excessively delay in pursuing common sense solutions.

Erosion Control. The loss of soil remains a critical resource management concern in virtually all regions of the state where the landscape is characterized by short, steep slopes or longer, gentle slopes.

It is extremely fortunate that Wisconsin has been in the forefront of the soil conservation movement for several decades. Scientists in the university system have contributed to the development of new conservation systems. The state's farmland preservation program remains one of the most effective in the country, and the state has been highly innovative in the formulation of local approaches to resource stewardship.

Hopefully, the commitment to land stewardship will never lapse because the potential for extremely high rates of erosion will always exist on a considerable portion of the state's cropland. Some shallow soils could be severely impaired in just a few years of bad management. Such soils often have less than 4 inches of topsoil remaining, and could lose 50 to 75 tons per acre per year, or up to one-half inch of soil. At such rates of erosion, topsoil could be lost in a matter of 5 to 10 years. Such land might not support profitable levels of crop or forage production for many, many years.

One of the most important challenges is determining how to integrate soil erosion control practices and water quality BMPs in profitable farming systems, *that also* comply with local, state, and federal land use, farm program, and water quality requirements. Herein lies a major challenge for social scientists, as well as academic leaders—tapping the full range of the system's scientific expertise needed to formulate, analyze, and implement the next generation of soil and water conservation policies. The university has not distinguished itself in this area to the degree it has in the physical sciences underlying effective conservation systems.

Forest. The forestlands of the state are clearly a major natural resource, and a key foundation for sustainable economic activity. Trade-offs among the multiple uses of forests, however, will pose major analytical and political challenges for the state. It is regrettable that in many states both public and private forestlands are abused or ignored because we lack acceptable institutional mechanisms to sort through the competing uses and public interests in forestlands.

Management of forests for multiple uses remains more an appealing concept than a practical reality. The difficulty of balancing timber and pulp production, wildlife, fisheries, recreation, mining, and watershed management demands on forestlands will grow even more complex in the years ahead. Major economic gains would no doubt follow successes in Wisconsin in finding a better balance among competing demands on forestlands, and then getting on with the task of actively managing forestlands to maximize the benefits that can be derived from their use.

What role will the university system play in establishing a process for society to sort through these conflicting demands? Where should the university place its emphasis—creating knowledge, training more individuals, or fashioning new public processes dedicated to resolving such conflicts? How can the university act upon its decisions regarding its role in an era of fiscal constraints and competing priorities?

Many other important natural resource issues warrant brief note. Climatic change and acid rain may be threatening the productivity of Wisconsin's forestlands. The impact of current forestry practices on soil productivity also must be closely monitored.

Wildlife. Fish and wildlife populations throughout the state enrich the lives of all citizens, and provide valuable recreational opportunities. The loss of habitat, water pollution, excessive hunting and fishing pressure, snowmobiles, and chemical pollution continue to place certain populations in jeopardy. The source of these pressures must be monitored and more fully understood before realistic remedial actions can be considered. Then, ways must be found to implement reasoned natural resource and wildlife management plans. Again, this last step is likely to be the most difficult, but is the linchpin if the public's initial investment in knowledge is to deliver on its promise of a better world.

Land Use and Cropping Systems. A critical challenge in sustaining the productivity of Wisconsin's natural resources is more carefully matching land use to the physical characteristics and limitations of the land. Indeed, a shift in land use can often help control erosion, protect water quality, sustain soil productivity, improve farm and forest profitability, and diversify habitat for wildlife. In much of Wisconsin, an ideal three- to five-year rotation on cultivated cropland will include up to two years of corn (not in back-to-back years), a year of soybeans, perhaps a year of small grains, and one or two years of a legume forage crop.

The prospect of important environmental and economic benefits following adoption of more diverse crop rotations received considerable emphasis in our recently released report *Alternative Agriculture*. These benefits result from several factors, well-documented over the years by research here in Wisconsin, as well as several other states.

The concept of crop rotation is foreign in parts of American agriculture, but not here in Wisconsin. Two factors, though, stand in the way of more thorough exploitation of the potential biological benefits of rotations.

First, changes are needed in federal commodity price support programs to encourage—rather than penalize—rotations. Second, new varieties of forages are needed, as well as new technologies and systems for improving forage yields and capturing their full range of benefits. Conservation management systems across the country will be based on reintegrating forages into cropping systems. Given the diverse opportunities for marketing and utilizing forages in the state, and the prospect of federal policy changes to provide new incentives for longer rotations that include grass or legume forage crops, the university should reassess all aspects of its research program on forages, from plant breeding to stand establishment, fertility management, and harvest technology.

Food Safety and Quality

Contrary to a view often expressed by agricultural and food industry leaders, there is some justification for most commonly expressed consumer concerns about the quality and safety of the food supply. While relative risks are often misunderstood, many consumers are reasonably well aware of the actual and potential risks in the food supply. Most people can remember their last case of foodborne illness, and are fearful that animal drug and pesticide residues are not as benign as some claim. A steady stream of new science keeps reinforcing this fear. Consumers are, in general, much more risk averse than regulators or polluters. Politicians are the only important group in society that are generally more risk averse than consumers, which goes a long way toward explaining the current crisis we face in public decision-making.

I am convinced that emerging knowledge and better science will intensify consumer concerns and lead to even stronger political pressure for stricter, more aggressive public and private sector efforts to provide safer, more nutritious, and more satisfying foods. Politicians interested in re-election and companies interested in staying in business will respond. Accordingly, major challenges lie ahead in the food sciences, the field of microbiology, and in the engineering disciplines key to the development of improved quality control procedures. I offer no bright ideas on how to restore reason in the political process.

Within the Beltway surrounding Washington, D.C., food safety experts in recent years have grown deeply concerned about microbiological contaminants in food. Food and Drug Administration officials have been trying to counter growing concern about pesticide residues in food (and pressures to redirect FDA budget resources) by issuing increasingly scary pronouncements about the threat of microbiological contaminants.

A leading scientist recently opened a symposium on food safety by proclaiming that the American consumer should now assume that virtually every piece of fresh meat, poultry, or fish is contaminated with some pathogenic, bacterial, viral, or chemical contaminant. Each year the Center for Disease Control keeps increasing its estimate of cases of foodborne illness, and is currently tracking the spread of a number of extremely troubling new bacteria and pathogens which appear capable of inflicting much more serious harm than other more common and better-known contaminants. Current science supports an estimate that over 6,000 people die each year from complications arising from foodborne illness. Many experts privately acknowledge the true number is probably far higher. The growing prevalence of antibiotic resistant strains of bacteria remains a troubling reminder that animal care practices can markedly impact food safety and human health, despite the 1989 Institute of Medicine (IOM) report which concluded that only 60 deaths per year can be traced to the subtherapeutic use of antibiotics. Again, the IOM committee acknowledges that the actual number may be much higher, and that many other serious health complications, short of death, probably result from subtherapeutic antibiotic use.

Food safety problems will prove most pressing in the dairy and livestock industries. Accordingly, states like Wisconsin that are heavily dependent on livestock agriculture will have to confront head-on the challenges of improving food safety. A multifaceted approach will be essential, including:

- monitoring the presence, virulence, and epidemiology of bacteria and other contaminants;
- altering animal feed manufacturing, distribution, and storage practices to slow down the spread of bacteria and pathogens among animals on a farm or in a feedlot, and from farm-to-farm or lot-to-lot;
- improving methods to detect and isolate animals, herds, or products that contain harmful bacteria or other pathogens;
- developing new quality control procedures in dairy processing plants, as well as in slaughter houses and meat processing plants; and
- educating workers on how to recognize and avoid potential hazards at all stages of the food system including production, processing, preparation, and retail sectors.

As a last line of defense, a major public education campaign is clearly warranted to teach a new generation basic skills and knowledge about how to handle foods. Too many consumers just are unaware that certain foods may cause illness, and must be properly handled and prepared. Regrettably, they have taken too literally assurances that the food supply is safe.

Promoting Health Through Dietary Change. Another major challenge is promoting improved health through dietary modification. Our 1987 report *Designing Foods: Animal Product Options in the Marketplace* highlighted a number of attractive opportunities to improve the nutritional attributes of animal products. Encouraging progress is being made in all sectors, but more clearly needs to be done. Moreover, scientists must turn their attention to a number of targets in addition to reduction in fat, cholesterol, and sodium content. Even in this era of plenty, millions of people in certain population subgroups consume too little calcium, too little available iron, and not enough of certain vitamins and minerals.

Exciting opportunities are emerging to modify the nutritional composition of foods through genetic engineering, classical breeding, altered management practices, hormonal supplements, and other practices. The so-called "fake fats" are likely to gain approval in the early 1990s, and could rapidly change how many common foods are manufactured and prepared. Estimates of the probable market penetration of Simplese and Olestra, if achieved, will result collectively in at least a 5% average decrease in the percent of calories from fat across the entire population. *On average* this could bring the entire population very close to the American Heart Association/Public Health Service/National Research Council recommended goal of no more than 30% of calories from fat. That in and of itself would be a remarkable accomplishment of profound significance in terms of improved health status.

The dairy industry will be another key target for major change as the nation strives to reduce consumption of fat and cholesterol. The industry has two options it could pursue in contributing to a reduction in the quantity of excess fat available in the food supply. It can cut overall production, or alternatively, reduce the average percent of butterfat in each hundredweight of milk. For a variety of reasons, the second option is likely to emerge as the most popular, particularly in a state like Wisconsin which is so well-suited for highly efficient dairy production.

For decades the pricing of dairy products has been based on a 3.4% butterfat standard. Analysts are now exploring a range of alternative standards for the pricing of milk. Three options are encountered in discussions with industry leaders:

- a lower butterfat standard, possibly coupled with a reduced price for milk that has more fat than called for by the standard;
- a new standard based on a minimum protein content; and
- a hybrid standard involving the ratio of protein to fat content, with premium prices paid for milk with higher ratios.

Protecting Sensitive Populations. In the early 1980s, federal regulatory officials in charge of the pesticide program ushered in a number of scientific refinements in the estimation of dietary risks. The basic goal was to lessen pressure for regulatory actions based on inflated estimates of pesticide risks. New methods and assumptions were incorporated into the risk assessment process designed to estimate risks more realistically in light of the actual extent of use of a pesticide, and the actual level of residues remaining in foods as consumed.

After years of divisive debate on the appropriate scientific basis for risk assessment, there is now a growing consensus that such an approach is both feasible scientifically and appropriate as a matter of public policy. In recent years, great strides have been made in estimating exposure to pesticides, food additives, drugs, and environmental contaminants among distinct population subgroups. Moreover, developments in toxicology, immunology, AIDS research, and other fields of science are beginning to shed light on the unique sensitivity of certain population groups to health hazards associated with exposure to infectious agents, microbiological contaminants, or chemical residues in food, water, or the air.

The most recent regulatory statutes have been drafted on the federal level: the reauthorization of the Clean Water Act, the reauthorization of the Clean Air Act, and the pesticide legislation. Most of these bills also have provisions that the EPA should regulate risk so that the most sensitive population group, or the most exposed population group, or both, are protected.

During the 1980's, Congress set regulations at levels that protected the average person. However, evidence is mounting that pregnant women, infants, and individuals with compromised immune systems or other health problems are likely to be much more sensitive, or that they are likely to be exposed at a higher rate, than healthy adults to many contaminants. Therefore, I think that as scientific information becomes available, legislation in the 1990s will reduce the risk level for people who are uniquely sensitive or vulnerable to various diseases.

Since the combination of higher rates of exposure and enhanced sensitivity result in much higher estimates of risk for certain populations, major scientific and regulatory policy questions lie just beyond the horizon. They will no doubt emerge as one of the major public health challenges for the next generation of scientists. The possible consequences for agriculture and the food system are, of course, of monumental significance, and will become a major focus of intellectual activity in several departments in several colleges on campuses across the university system.

Social Expectations
and Economic Performance

There has been relatively little attention to farm structure issues since the late 1970s. The collapse of many rural communities in the 1980s, and more recently, the Bovine Somatotropin (BST) debate have rekindled what may become a spirited national debate about the structure of agriculture. Broad-based interest in sustainable agriculture and national dairy policy reform will also help reframe longstanding concerns about size of farms and the resiliency of a family based system of agriculture.

A debate about the structure of agriculture is clearly under way here in Wisconsin. It is an important debate, grounded in deeply-felt opinions about the virtues of family farm agriculture. As a nation, we have felt strongly enough about the family farm to invest $170 billion during the 1980s in an attempt to maintain its economic viability. Support for the family farm is also one of the stated purposes of all domestic farm programs. This is of great significance for Wisconsin because the agricultural sector remains as close to a family farm structure of agriculture as any state in the nation. As federal farm policy targets a larger portion of public dollars to family farmers, Wisconsin agriculture will benefit disproportionately relative to other states.

The BST debate may finally force the nation to confront the profound forces in rural America arising from the interactions of new technology and agriculture policy. Many agricultural industry leaders in Wisconsin believe that dairy farmers must have access to all technologies that promise to improve productivity. BST is expected to do so rather dramatically by increasing milk production per cow some 10 to 15% on average, while increasing feed efficiency.

Many dairy farmers are opposed to the adoption of BST, at least under current dairy policies. The reason is rather obvious. Virtually all credible studies of the consequences of BST adoption envision at least a 5% increase in the national supply of milk. Under current policy, such an increase in supply—in the absence of comparable growth in demand (very unlikely)—will result in at least one, and possibly two 50-cent cuts in the price of milk. (Under current policy, price cuts are triggered automatically by large surplus purchases by the government.)

Accordingly, if BST is widely adopted, dairy farmers must not just recover the cost of BST, but must also learn to live with a 50-cent or $1.00 cut in income per hundredweight, whether they use BST or not. Given the narrow margins on most dairy farms today, enthusiasm for BST wanes as the full potential consequences of this technology are contemplated. Is it realistic for society to ask dairy farmers to make such a sacrifice in order to keep up the rate of technological progress in the sector?

The potential economic benefits from adoption of BST cited by proponents of the technology could be realized more fully under a different national dairy policy. The key change would be the adoption of a production-based quota system, so that each farmer could meet his or her quota with perhaps 10% fewer cows, and presumably lower costs. There are, however, a number of complex political hurdles that stand in the way of such a quota system. Such a change in policy is likely to introduce other sources of inefficiency in resource use patterns. The political hurdles include disagreement over the appropriate regional distribution of quota, how to allocate quota to farms within a region, and how to set and enforce prices.

BST is but one of the several examples of serious conflicts in the nation's agricultural, conservation, and technology policies, conflicts which are growing rapidly in importance, and simply must be confronted by the Congress in shaping future farm bills.

At the state level, another issue is agricultural development in remote regions. My sense is that cropland in the southern two-thirds of the state is, with limited exceptions, fully utilized, or nearly so. In the northern third of the state much of the cropland is idle. In the late 1800s, some northern counties were among the country's most prosperous and productive agricultural regions. I am convinced that agriculture can and will eventually rebound in northern Wisconsin. Profitable farming enterprises will be based on forages and beef cattle; dairying, to some extent; the production of grass and legume forages and seed; and speciality crops including berries, apples, honey, maple syrup, wild rice, and a diversity of other commodities. What role will the university play in facilitating agriculture-based, rural economic development in northern Wisconsin?

Animal Welfare. Manure management and the role of technology are already pressing contemporary issues for livestock agriculture. A wide range of animal welfare issues are bound to crystallize in the 1990s, greatly complicating the political agenda confronting livestock industries. Because of the scale and type of livestock farms in Wisconsin, the emergence of animal welfare as a force shaping national policy may paradoxically create economic opportunities for Wisconsin agriculture.

The major targets of animal welfare groups are likely to be capital and drug-intensive confinement operations that severely restrict the movement of animals. The environment in some of the buildings used in such operations—the air, water, access to sunlight, exposure to infectious agents—can be very unhealthy, inflicting unnecessary pain and stress on animals and requiring more frequent reliance on drug therapies to ward off disease and maintain acceptable rates of growth. These concerns are bound to weave into the food safety, international trade, sustainable agriculture, and family farm agendas.

Who Will Pay to Improve Environmental Performance? It remains to be seen how society will share the costs and distribute the benefits of new policies dedicated to improving the environmental performance of American agriculture. Several major nuts and bolts questions about money will have to be resolved in the 1990s.

While federal taxpayers are underwriting the major new soil erosion control provisions in the 1985 farm bill, it is unclear whether the federal or state governments, or land managers will pick up the tab for new water quality initiatives expected in the 1990 (or 1991) farm bill. Fiscal pressures at the federal level may dictate a largely regulatory approach, or may require building water quality goals into future commodity programs as a way to partially redirect an existing piece of the fiscal pie. States wishing to pursue a more gentle, voluntary, cost-sharing approach will no doubt be urged on by federal agencies, but at their own expense.

An even more basic question will confront policy makers in the 1990s. The transition to sustainable cropping and livestock systems will be information and management intensive. Who will help transfer new knowledge and skills to farmers, and how will their efforts be paid for?

Because of severe fiscal limitations in the face of a growing list of partially met program responsibilities, I cannot envision a predominant role for cooperative extension in this process. Even a minimally realistic effort to support sustainable agriculture would necessitate a multifold increase in both the federal and state fiscal commitments to extension. Such an increase is unlikely. As a result, a variety of ways must be found for the private sector to take on a major role in this process. The key for such a private sector role to emerge will be expanding the acceptance and effectiveness of private consultants and other specialists that can help farmers with highly specific and technical management problems. Another key step will be devising ways for co-ops and agribusiness to sell services like soil fertility maintenance and crop protection, in contrast to basing income prospects solely on the gross volume of fertilizer or pesticides sold.

INSTITUTIONAL STRENGTHS AND WEAKNESSES: OPPORTUNITIES AND IMPERATIVES

The University of Wisconsin System has major strengths to draw upon in meeting the challenges that lie ahead. I foresee many opportunities

===
Table 3.1

Reliance on Federal and State Funding at Selected State
Agricultural Experiment Stations, FY 1988

	USDA Competitive Grants	USDA Special Grants	Other Federal Competitive Grants*	State Appropriations
Wisconsin	2,486,231	224,981	13,384,000	21,329,000
Minnesota	1,275,583	578,464	2,436,000	33,554,000
Iowa	571,550	781,346	1,561,000	18,630,000
California	6,351,327	2,716,305	23,696,000	89,262,000
New York	3,093,150	995,083	16,903,000	37,560,000
North Dakota	100,000	2,976,733	561,000	11,636,000
Mississippi	106,243	6,530,817	995,000	17,063,000

===
*Principally NIH and NSF competitive grants.

Source: Tables A.14 and A.15, Investing in Research, Board on Agriculture,
National Research Council, 1989.

to build upon the system's strengths in tackling some of the system's current weaknesses and shortcomings.

System Strengths

The University of Wisconsin System is large, diverse, and well-endowed with capital infrastructure. It enjoys strong support throughout the state, and can expect the help of the legislature as new needs emerge.

The UW–Madison campus is one of the nation's great research universities. Its scientific accomplishments are known around the world, and the intellectual environment rivals that found on the best universities anywhere.

The College of Agricultural and Life Sciences (CALS) at UW–Madison has compiled a superb record of scientific achievement. On a per scientist basis, the ability of the College is unmatched in winning both U.S. Department of Agriculture grants and competitive grants awarded by other federal agencies, principally the National Science Foundation (NSF) and the National Institutes of Health (NIH), as evident in Table 3.1. CALS has also enjoyed strong support from the state's principal farm organizations and agribusiness organizations. Indeed, campuses throughout the UW System have active and supportive relationships with all key constituent groups in the state.

Weaknesses and Shortcomings

Based on my admittedly limited review of the institutions within the system, there appears to be at least one, possibly a few campus programs that are difficult to justify. Some programs are very thin, and offer students educational opportunities simply not up to the standards available elsewhere in the system. Even if the resources were available to improve these programs, it is questionable whether demand from the prospective student population really justifies such investments.

Several observers note a series of actions that have been taken to strengthen one institution or program, but only at the expense of other institutions or programs. Such actions may do little to strengthen the system as a whole.

On the UW–Madison campus, the interest, willingness, and ability of faculty to undertake multidisciplinary research and education activities is suspect. The culture of scientific achievement on campus is firmly embedded in individual investigator research. Collaboration among biological and physical scientists in related fields has produced important breakthroughs, and is not uncommon. But effective collaboration across unrelated disciplines—particularly among biological and social scientists—appears regrettably rare. The ability of faculty to work across college boundaries is even more limited.

This problem in taking on research and education activities that necessitate a multidisciplinary approach lies at the heart of many controversies surrounding CALS programs and priorities. It will become an even more serious problem in meeting the challenges of the future, which will reflect social concerns that can only be understood and resolved in a multidisciplinary context.

Regrettably, despite much recent effort throughout the system, the appropriate role for UW–Extension, and how it will meet its responsibilities, warrants further reflection. The organizational changes already adopted are only part of the problem, and can be at best only a partial solution. At the federal level, the role of extension is under review, and clearly may be subject to significant change. The role of extension in advancing adoption of water quality BMPs and sustainable production systems will no doubt evolve in very different ways across the country.

In recent years many leading faculty and administrators on the UW–Madison campus have become embroiled in divisive public debates over BST, the university's role in support of sustainable agriculture, and other public controversies. As a result many people around the state, and even on campus, have lost perspective of the overall achievements of the college, and grown either confused or wary about its goals. These debates have had an impact on activities within the college, as well as in the college's relationships across campus. In any dynamic institution some tension over control of budgets, turf, and program direction is healthy. At some point, tension becomes counterproductive if not channeled into constructive dialogue and interaction. For example, a much more mutually supportive relationship should be forged between the college and the Institute for Environmental Studies.

Last, many faculty members question the policies now governing the receipt and use of private sector funds in support of research on the UW–Madison campus. They worry that the overall cost to the university— controversy over research direction, credibility of results, restrictions on academic freedom—may exceed the benefits of this source of support. This touchy issue is coming to a head on a number of land grant universities across the country.

Opportunities and Imperatives

Attractive opportunities exist for the university system to build on its strengths in addressing some weaknesses. Grantsmanship skills on the Madison campus are highly developed as indicated by the rate of success in securing individual investigator grants. These skills can form a basis upon which to cultivate improved grantsmanship skills in pursuit of multidisciplinary research funds. Assuming favorable action on the National Research Council's Board on Agriculture's proposed $500 million national research initiative, two new categories of multidisciplinary grants will be offered through a markedly expanded USDA competitive grants program. Clear and tangible incentives for faculty interested in developing multidisciplinary proposals could accelerate progress on campus in attracting a significant increase in competitively awarded grants in support of multidisciplinary research.

University researchers and administrators may need to work out cooperative relations with other state agencies handling natural resource and agricultural policy. Recently I had an interesting series of exchanges at a conference on water quality. After the conference, I accidentally overheard a conversation between a state legislator, who chaired an environmental committee, a member of an environmental conservation organization, and a program leader from a state department of agriculture. (This was a particularly interesting trio because during the previous legislative session in this state, the state legislature had passed a new groundwater bill that was implemented by the environmental conservation organization and administered by the agriculture department). The environmental activist told the legislator how important it was for the state to move its groundwater research. The question arose as to the role of the university in this research. The state agriculture department representative observed that the university can be an expensive place to put resources to address groundwater research. He went on to say, that the university has a 40% overhead rate and is too submerged in the "good old boy" network. Finally, he stated that he thought at the university only ten cents of every dollar resulted in new thinking.

I believe that conversations like this one are going on in many states. Clearly, universities must begin to identify their advantages. Although the University of Wisconsin has done some significant and important assessments of its programs, issues remain that must be looked at, because many natural resource problems must be solved one part at a time. One cannot legislate from Washington, or even Madison, how farmers can take care of their natural resources. The majority of farmers will need to incorporate new kinds of resource conservation and pest control into their farming operations. They cannot do this without research.

Controversy over the influence of private sector research could be alleviated by changing the conditions under which such funding is received.

A model for how this could be done already exists on campus. Back in the 1950s when most corn hybrid seed yield trials at Midwestern land-grant universities were supported by the seed companies developing the new varieties, questions kept arising about the reliability of the results and fairness of experiment station recommendations. To solve this problem, experiment station leaders decided to take over the funding of yield trials. At the same time, academic leaders sought general support from the seed companies to finance fellowships, equipment, and other expenses. Many companies responded favorably to the request. For nearly three decades, this arrangement has worked well for plant breeders. Why shouldn't private sector support for herbicide and other pesticide trials, or animal drug experiments be granted to the university under the same or similar conditions?[1]

SUMMARY

Cultural and intellectual diversity has always been a hallmark of the University of Wisconsin. Emerging challenges faced by citizens in the state will press the university to rethink some of the principles and processes governing the basic missions of the university. This process of reassessment is already well under way throughout the system.

The university exercises a degree of control over the supply of educational experiences it wishes to offer. But what about the customer, the young people in the state who are interested in pursuing higher education? Why do they enroll in college? What do they hope to achieve?

For how many students is college a stepping stone toward a well thought-out professional career? For how many is it an opportunity to learn more about the world and matters of personal interest, while future career opportunities are explored? Is the university currently offering students the breadth of education and intellectual challenge they are seeking?

Hopefully in the course of this strategic assessment of university activities in agriculture and natural resource management, opportunities will arise to assess these important questions about the demand for educational resources by interacting with a cross-section of students. Because the rate of change in society has accelerated greatly in so many areas, my guess is that many students are seeking an educational experience that differs in some important ways from what many departments in the university have offered in the past. Developing ways to meet emerging aspirations of students while also responding to the many challenges that lie ahead will keep things lively within the university system for many decades to come.

NOTES

1. The genesis of this idea can be traced to a conversation with Dr. John Axtel of Purdue University, who recently described for me the history of private sector support for corn hybrid yield trials.

4

The International Economy:
The Next Thirty Years

M. ANN TUTWILER

INTRODUCTION

Some might think it crazy to ask a thirty-one year old to speculate on where the international economy is heading in the next thirty years. I can't remember where I was when Kennedy was shot; I was only 10 when the Grateful Dead played Woodstock; I can't recall a time before space travel. But, neither can I remember when the United States dominated the world economy—I was 13 when the United States suspended convertibility of the dollar; nor can I recall a time when the word international was not followed by interdependence. Since I was born, the nature of the global economy has changed, irretrievably. The nature of agriculture's relationship to the world economy has also changed, irretrievably. U.S. agriculture depends more on exports and purchased inputs than it did after World War II. More, and different countries buy U.S. agricultural products than previously; fewer, and more similar countries compete with U.S. agricultural products. International macroeconomic policies have as much to say about the income of U.S. farmers as does U.S. agricultural policy.

U.S. AGRICULTURE
AND THE WORLD ECONOMY

The link between U.S. agriculture and the global economy is not new. Since colonial times, U.S. agriculture has needed the international market for its livelihood. But, as world market growth and U.S. productivity increased in the last two decades, the nature of the U.S. dependence on the global economy has changed.

43

Exports. U.S. agriculture's dependence on exports has risen dramatically. By 1980, almost 40% of U.S. cropland was planted for export. Exports accounted for about one quarter of farm income (Kost and Jabara, *1985 Yearbook of Agriculture*, 1985). More than one million people work in jobs related to agricultural exports, and agricultural exports account for one-third to one-half of the income of 16 states. Finally, because the demand for food in the United States is almost stagnant, U.S. agriculture needs the foreign market simply to maintain the size of the sector.

While U.S. agriculture's increased reliance on exports mirrors the rest of the U.S. economy, the nature of agricultural production makes this dependence more precarious than it is for other economic sectors. Individual producers cannot control the prices they receive for their commodities. When sharp currency swings turn exchange rates, and thus international agricultural prices, from favorable to unfavorable overnight, agricultural producers are doubly exposed.

New Buyers. As the importance of U.S. agricultural exports was rising, their destination was changing. Prior to World War II, there were three net importing regions—Western Europe, China and Japan. All other countries and regions were net exporters. After World War II, this pattern changed. The number of net exporters declined, while the number of importers increased. By 1985, most Lesser Developed Countries (LDCs) and centrally planned economies were significant importers, and Japan remains as the only developed country importer (McCalla, 1989).

The change in destination has been particularly dramatic for the United States. (The following figures are for wheat). In the early 1960s, 20% of U.S. exports went to other developed countries; now only 13% goes to these countries. LDCs took almost 60% of U.S. exports in the early 1960s under Public Law 480; now, LDCs buy half of U.S. exports, mostly under commercial terms. Exports to centrally planned economies have risen from 10 to 30% over the same period (McCalla, 1989).

The European Community, followed by Japan, still account for the largest markets for the United States. But, the demand for food in these relatively rich countries grows slowly. The potential for rapid increases in food demand abound in the United States' new markets. However, developing and centrally planned economies have been plagued by low income growth and erratic domestic production. This makes them more volatile and more fragile markets for the United States.

Inputs. In addition to being more dependent on exports, agriculture is also more dependent on purchased inputs than in the past. Typically, the cost of agricultural inputs are set in markets that are less competitive than agriculture, and that do not necessarily respond to changes in agricultural prices. More importantly for this discussion, the price of many of these purchased inputs is determined by forces in the international market.

With the collapse of the system of fixed exchange rates, the rise in oil prices, and the development of the Eurocurrency market, the world became one huge capital supermarket. Farmers in Topeka had to compete with carmakers in Tokyo for funds. This unified capital market benefited farmers in the 1970s, when farmers borrowed to expand their production in the face of rising demand. But, when a confluence of factors drove up world interest rates in the early 1980s, farmers' borrowing costs rose with them. The farm debt crisis followed.

The OPEC oil cartel also has had dramatic effects on production costs. The share of energy related inputs in producers' total costs doubled between 1972 and 1981 (USDA, 1983). Energy prices have moderated in the last several years, but only because of soft international demand and economic pressures on OPEC members to produce above their quotas. Energy costs are still determined largely by forces outside the control of the United States. Finally, the global policies that created high inflation in the 1970s and deflation in the 1980s has caused dramatic swings in the price of the farmers' most important input: land.

Competition. The face of the competition has also changed. Except for Argentina, all of our major agricultural competitors are other developed countries. All, to greater and lesser degrees, subsidize their agriculture. In the European Community and Canada, government and consumer transfers added roughly 50% to the value of agricultural production in 1986. Even in the less subsidized countries of Australia and New Zealand, transfers add 13% to the value of farmers' production (For comparison, U.S. transfers added 36% to the value of farmer's production) (Blandford, de Gorter, Dixit and Magiera, 1988). This means that farmers are competing not just against other countries' farmers, but also against other countries' budgets. It also means that U.S. competitiveness is not simply determined by lower costs, it is also affected by national policies in Europe, Canada, Australia and elsewhere.

Macroeconomic Policies. The tremendous upswing in agricultural exports of the 1970s and the downturn in the 1980s can be almost wholly attributed to global macroeconomic factors. In 1973, U.S. exports jumped $5.7 billion (in 1985 dollars) and continued to climb by $2.1 billion annually (five times the historical rate) throughout the decade (Rossmiller, 1987). The jump in 1973 can be explained by the Soviet Union's "Great Grain Robbery" and by numerous failed harvests, but the continued expansion was due primarily to high global growth rates caused by expansive macroeconomic policies enacted in the wake of the OPEC oil embargo.

In 1981, U.S. exports peaked. The following year, they plummeted $4.7 billion and began falling at an annual rate of $1.6 billion until 1986 when exports turned around. The 1986 turn around coincides with a fall in the value of the dollar the previous year. The steep climb in exports, the fall, and the rebound can all largely be explained by changes in international macroeconomic conditions and by international exchange rates.

Gyrations in macroeconomic policies have generated gyrations in exchange rates that have had a critical impact on U.S. agriculture. Between 1970 and 1979, the index of the trade-weighted value of the dollar fell from 125% to 75% and the nominal value of U.S. agricultural exports jumped from $7 billion to roughly $35 billion. While exports peaked at $43 billion in 1981, the dollar began climbing. In 1982, exports fell from that peak to $29 billion (Economic Report of the President, 1987). It is estimated that the appreciation of the dollar in 1980–81 cost U.S. agriculture roughly 10% of its export earnings (Longmire and Morey, ERS, Foreign Agricultural Economic Report, No. 193, 1983).

Nothing is the Same. Virtually nothing is as it was for U.S. agriculture. Exports contribute a significant portion of farm income; prices of inputs are determined in the world market; U.S. markets are more diverse and more unstable; U.S. farmers compete with foreign governments as well as foreign farmers.

What matters—and more importantly—what will matter to U.S. agriculture in the future are the international financial regime, international trade system, global economic growth, and global agricultural production. In the following pages, I will talk about what is driving these now, and what I expect to be driving them in the future.

THE INTERNATIONAL FINANCIAL REGIME

The Last Thirty Years

Initially international capital flows served to finance trade and facilitate foreign investment. Now, international capital flows tower over international trade. In 1984, the total value of world capital flows was estimated at $30 trillion (Garten, 1985). By comparison the value of world trade in that year was $2 trillion, and the value of world agricultural trade a mere $220 billion (FAO, 1986). While no single event elevated international finance to its position of dominance, the collapse of the Bretton Woods system and the OPEC oil embargo in 1973 heralded the change. However, the foundations of the hegemony of international finance really began in the last 1950s with the emergence of the Eurocurrency market.

The Eurocurrency Market

The Eurocurrency market was born in the late 1950s when the Soviet Union began depositing its dollar earnings into European banks instead of U.S. banks where they feared these deposits might be confiscated. International banks began using the Eurocurrency market to avoid domestic interest rate ceilings and reserve requirements. Corporations also use the Eurocurrency markets to buy and sell currencies needed to complete daily transactions. The market expanded rapidly, growing from $7 to $250 billion between 1963 and 1975 (Karlik, 1981). The Eurocurrency market facilitated large, destabilizing capital flows which ultimately forced the collapse of the Bretton Wood system. The Euro-currency market also enabled the international financial system to smoothly recycle the massive influx of petrodollars after the first oil price shock.

The Collapse of the
Bretton Woods System

Under the Bretton Woods system, the United States was committed to exchanging dollars for gold at $35 per ounce. But, by the early 1970s, officially held foreign dollar reserves began to exceed U.S. gold reserves and financiers began to doubt the ability of the United States to honor its commitment. Their concerns were aggravated by the large U.S. balance of payments deficits in 1970 and 1971 which increased the number of dollars held overseas. Financiers began selling dollars in the Eurocurrency markets, anticipating that the United States would devalue the dollar. Their speculation became self-fulfilling, and the dollar was devalued and its convertibility to gold suspended. Several attempts were made to re-establish fixed exchange rates, but all failed.

Dirty Floating

The Bretton Woods system was replaced by a regime of "dirty floating." Under this regime, governments allow exchange rates to freely float and find their market levels, but occasionally intervene in the foreign exchange markets to prevent wide swings in their currencies' value. Financiers and speculators also trade currency on the Eurocurrency markets in anticipation of exchange rate movements. According to a recent survey, in 1984, nearly $150 billion was traded each day on world foreign exchange markets, an amount almost equal to total annual agricultural trade (Group of Thirty, 1984).

When such sums shift in response to speculative pressures or higher yields, they can contribute to changes in exchange rates, and can frustrate central banks trying to influence exchange rates. For example, after the September, 1989 Group of Seven meetings, central banks began selling dollars in the foreign exchange markets to try and depress the dollar exchange rate.

Traders, however, felt that the dollar had been pushed to an artificially low level by central bank sales, and purchased dollars. Despite central banks' coordinated, and expensive efforts, the market perception of the dollar's value prevailed. Eventually governments may be forced to change their interest rates in order to lower the dollar's value (Donald, 1989).

The Rising Yen

After the dollar-based Bretton Woods system collapsed, it was replaced by a multi-currency reserve system. Central banks needed to hold currencies that paralleled the patterns already developed by private traders and investors. While the dollar still anchors this system, the yen, the deutschemark, the pound sterling, the French and Swiss francs have grown in importance.

The multiple reserve currency system subjects the macroeconomic policies of the reserve countries to increased international scrutiny. Under the Bretton Woods system, the United States could conceivably run large balance of payments deficits with no adverse effect on the value of the dollar. Under a multi-currency reserve system, no country can sustain a deficit for long without risking that traders might move to other, stronger currencies (Roosa, 1982). Thus, the availability of alternative reserve currencies increases the exposure of each to the risk that other central banks, and financiers, will shift holdings among them—often through the Eurocurrency markets—in response to speculative and yield pressures, as well as in concern about the wisdom of a country's macroeconomic policies. In particular, this move away from dollars also increases the scrutiny accorded U.S. macroeconomic policies.

Volatility. Floating rates have also increased volatility in the foreign exchange markets. One illustration: since 1971, the dollar fell 35% against the deutschemark for two and a half years, oscillated between plus and minus 10% over the next two and a half years, fell 35% over the next four years, rose 99% over the next five years, and then fell 24% in the seven months between February 1985 and August 1985 (Marris, 1987).

Whether, and how much, this volatility matters is difficult to determine. There is some evidence that it does matter, quite a bit. Between 1976 and 1980, central banks deemed stability important enough to have spent about $100 billion to dampen volatility in the foreign exchange markets. And, while slow trade growth has many causes, since the collapse of Bretton Woods, the rate of growth in trade fell from 8.5% in the 1960s to 4.6% in the last half of the 1970s (Roosa, 1982).

The Next Thirty Years

There is great nostalgia for fixed exchange rates. Floating exchange rates have complicated life for agricultural producers and policymakers alike, in large part because the importance of exchange rates has only recently been recognized in agricultural circles, and has not fully been accounted for in agricultural policy.

Is it possible to return to a fixed exchange rate regime? The short answer, in my opinion, is no. There are several reasons. First, a fixed exchange rate regime would require one country or a collection of countries, to subordinate its (their) individual economic goals to the global good (Feldstein, 1987; Uchitelle, 1988; McCracken, 1987). For example, the United States has been unable (or unwilling) to reduce its budget deficit and thereby ease pressure on world interest rates, despite demands from our allies. While such selflessness is unlikely in any case, it would be particularly difficult if a prolonged period of independence is followed by interdependence. The fact that coordination would require the acquiescence of three major powers (the United States, West Germany and Japan) instead of the discretion of one power further complicates efforts at coordination.

Each country has its own policy preferences. West Germany, forty years after World War II, is still paranoid of inflation and is unwilling to tolerate rates that to the United States would seem tame. Japan has been unable (or unwilling) to reduce its domestic savings rate despite international demands. Moreover, no country would raise interest rates prior to a presidential election, regardless of international considerations. The difficulties of establishing the European Monetary System are a testimony to the difficulties of coordination. Even among countries integrated in a strong and durable trading community, where the benefits of stable currencies are incontrovertible, a coordinated exchange rate system has met with resistance.

Second, the financial markets will test a fixed exchange rate system any time the political, economic or speculative conditions warrant. As mentioned earlier, the central banks' drive to lower the value of the dollar in late September was frustrated because traders felt that the United States and West Germany needed, respectively, to lower and raise their interest rates.

Third, pegged systems have a self-defeating tendency towards rigidity because governments can use an undervalued currency rate as a form of export subsidy. Resistance to changing exchange rates (which was prevalent even during the Bretton Woods regime) and eventually even a managed float can become artificially fixed (Artus and Young, 1980).

Fourth, and most important, no country (or institution) has emerged as world leader to exert a stabilizing influence on the global economy.

Recent U.S. macroeconomic policies have not inspired confidence in the United States ability to behave responsibly. Japan, whose economic strength is gradually dislodging the United States as the world's dominant economy, is asking for greater representation at the IMF, but is not yet ready to assume the political mantle of leadership (Uchitelle, 1988). Western Europe, despite 1992, is still too politically fragmented to fill a leadership role. The International Monetary Fund, whose Special Drawing Right was once held up as a possible international currency, does not have the political or economic wherewithal to run a fixed exchange rate system. (Much of the preceding discussion is taken from Tutwiler and Elliott, 1988.)

Implications for U.S. Agriculture

International finance will continue to dominate the ability of traditional commodity policy to support U.S. producers. Unrelated macroeconomic events—like the U.S. government's laissez faire attitude toward the value of the dollar in the early 1980s—will continue to frustrate the efforts of agricultural policymakers. Moreover, trying to override these macroeconomic phenomenon with price policies could get exorbitantly expensive.

Agricultural policymakers have several challenges. First, they must attempt to join in the debate on macroeconomic policy formulation. This will not be easy. Macroeconomic policymaking is even more rarified than agricultural policymaking. And, understanding macroeconomic relationships is perhaps even more difficult than understanding how U.S. commodity policy operates. Moreover, monetary policy is not made by democratic process. The Federal Reserve and the Treasury do not hold open hearings on interest rate and exchange rate policy. Fiscal policies relating to taxes and expenditures are made more openly, but often agriculture has competing stakes. Farmers do not wish to see cuts in deficiency payments, but a smaller budget deficit would bring a lower dollar and, with it, higher exports.

Second, agricultural policies need to be more flexible, and less dependent on price supports. The 1985 farm bill has made some steps in this direction by basing loan rates on market performance. But, it has not gone far enough. As we have seen, a rise in the dollar's value can change U.S. agriculture's competitiveness overnight. The current legislation changes prices only slowly, and may not take account of what changes in the exchange rate have done to competitiveness. Also, by fixing prices, U.S. agriculture could still provide super-incentives for foreign producers when the value of the dollar rises. U.S. agricultural policymakers need to begin exploring policies that take account of changes in macroeconomic conditions, without sacrificing U.S. farmers *or* U.S. taxpayers.

THE INTERNATIONAL TRADING SYSTEM

The Past Thirty Years

In the years following World War II, global trade grew dramatically. Between 1950 and 1986, it rose from $60 billion to over $2 trillion (IMF, 1987). Most economists attribute this astonishing increase to the freer markets fostered by the General Agreement on Trade and Tariffs. As world trade has increased, so has the United States dependence on both exports and imports. In 1984, approximately one-sixth of U.S. national income was related to traded goods and services (Garten, 1982).

Rising Protectionism. Overall, the expansion in world trade has benefited the United States by increasing national income, and by allowing consumers to purchase more and cheaper goods. But, certain sectors have suffered. About 70% of U.S. products are exposed to international competition. This competition has cost jobs in the textile, steel and automobile industries. These displaced workers have led the charge for increased protectionism. In many cases, the demand for more protection is simply rhetoric. However, eminent economists have noted an insidious increase in protectionism.

Since World War II, GATT has been extremely successful in reducing tariffs on manufactured goods. The Kennedy Round (1962–1967) reduced tariffs by an average of 35% on manufactured goods and 20% on agricultural products (Baldwin, 1981). The Tokyo Round, negotiated in the 1970s, reduced tariffs by another third.

But, while the use of tariffs has fallen, the use of non-tariff barriers (NTBs) increased. These NTBs are much less visible but can have an equally invidious effect on trade flows. This new form of protectionism has been difficult to quantify. Estimates of quantitative restrictions and voluntary restraint agreements suggest that approximately one-fifth of developed country manufactures were affected by non-tariff barriers in 1980—a four-fold increase since 1970. The frequency of quantitative restrictions under the emergency safeguard code (which is often used as a justification for long term protectionist measures) has risen from three cases between 1949 and 1958 to twenty-five between 1979 and 1986 (World Bank, 1987a).

There has also been an increase in process protectionism, of using quasi-judicial means to discourage imports. In addition, the scope of non-tariff barriers has expanded to include exports from developing countries (Ostry, 1987). Developed countries increasingly use NTBs to protect industries that face tough competition from developing countries, such as textiles, clothing, footwear and steel.

This new kind of protectionism is simply a beggar-thy-neighbor policy in sheep's clothing. Many of the mechanisms, which can include subsidies, voluntary restraint agreements, import quotas, standards, licensing agreements, and government procurement practices, appear to be legitimate efforts to maintain quality or redistribute domestic income. But, they succeed in choking off trade flows and are much more difficult to detect and to regulate.

Trade and Development. While protectionism appears to save jobs in some sectors in the short run, in the long run it will cost jobs in other areas. The less developed countries we exclude from our markets will be hard pressed to increase their purchases from us, if they cannot earn foreign exchange. U.S. agriculture, in particular, is vulnerable to this long term boomerang because it is so highly dependent for export growth on developing countries who are most hurt by protectionism. The increased protection has been aimed at restricting exports of labor intensive goods from precisely these countries. Such efforts to save manufacturing jobs in the developed countries will eventually cost jobs in agriculture.

By restricting market access, the developed countries deny the less developed countries an opportunity to earn crucially needed foreign exchange. Without that foreign exchange, heavily indebted countries are unable to service their debts. Many, such as Argentina and Brazil, have been forced to expand exports of feed and food grains in competition with U.S. grains, in attempt to earn the needed foreign currency. Others, such as Mexico, have been forced to restrict their imports, including agricultural imports, in order to conserve scarce foreign reserves. Not only have these import restrictions cost developed country jobs in the short run, they may also cost jobs in the long run, if developing countries cut back on imports needed to maintain domestic investment and productivity.

To the extent that trade barriers inhibit income growth in the less developed countries, the shift from a grain based to a protein based diet will be slowed. This side effect of protectionism is crucial for U.S. agriculture. As will be discussed more fully later, about half a billion people in the developing world are in the midst of this dietary transition (Sanderson and Mehra, 1988). Many of these people reside in the countries that suffer the most stringent restrictions on textiles and clothing, such as China, India, Indonesia and the Philippines.

Thus far, trade restrictions have not halted the increase in LDC manufacturing exports. As a proportion of total LDC exports, manufactures have risen from 10 to 65% since 1955 (Crook, 1989). But, despite their obvious advantage in consumer electronics, textiles, clothing, and footwear, less developed countries have only a small portion of the developed country market.

While it is difficult to state with certainty that increased trade restrictions have harmed developing or developed countries, it is important to point out that the adverse effects of trade restrictions may appear later, because investment in export sectors may have been inhibited. Also, the pattern of trade may have been altered, even though total trade has not diminished. Finally, in the absence of these measures, trade might have continued to expand faster than world income, instead of barely keeping pace as it has in the last decade (Corden, 1984). Moreover, as these countries continue to climb the development ladder, access to developed country markets will become increasingly important to their—and our—economic prosperity.

The Uruguay Round. As mentioned earlier, the United States' principle competitors and largest markets are now other developed countries. Almost half of U.S. exports are bought by other developed countries (FATUS, 1989). These countries all intervene in and protect their agricultural markets. This intervention has encouraged excess production which has had to be exported in competition with U.S. products. It has also closed off some important markets for the United States. The Uruguay Round of multilateral trade negotiations currently taking place under the auspices of the General Agreement on Tariffs and Trade (GATT) has focused on agricultural policy reform.

When the Uruguay Round began in 1986, world agricultural markets confronted huge surpluses, burgeoning budgets and costly trade wars. Estimates of the cost of developed country agricultural policies topped $200 billion (OECD, 1983). Since that time, the pressures that drove countries to the bargaining table have eased. The North American drought and policy changes in some countries have reduced the surpluses and the attendant budget costs. But the policies that led to the agricultural crisis still exist, the long term pressure on supply remains, and the multilateral stakes have not changed.

The Next Thirty Years

Over the next thirty years, the pressure for protection in the developed countries will increase as more LDCs climb the development ladder. As the economies of the newly industrialized countries (NICs: South Korea, Hong Kong, Taiwan and Singapore) continue to mature, they will begin competing with more sophisticated developed country products. Thus, the pressure for protection will spread to new industries.

A particular danger will be the threat of increased protectionism in these NICs, as lesser developed countries begin competing with their labor intensive products. Already some of the NICs, notably South Korea, are highly protectionist. This threat is compounded by the fact that LDCs are accorded special and differential treatment in the GATT which allows them to protect domestic industries. It is also compounded by the poor example set by the developed countries.

Unless GATT can avert it, most of this increased protection will likely be in non-tariff barriers, which are much more debilitating than tariffs because they are less visible and more unpredictable.

The next thirty years will be crucial ones for the GATT. If GATT succeeds in attacking agricultural support measures it will be able to cope with these challenges. If GATT fails, there will be a slowdown in the growth of global trade that will damage U.S. agriculture.

If GATT does succeed in agriculture, some forms of policies—such as export subsidies and import quotas—will be forbidden or at least their use will be strictly limited and controlled. If countries wish to operate import quotas, they will have to make minimum access commitments. Other forms of policies, such as deficiency payments and variable levies, will be phased out over time and will be replaced with less disruptive policies. Most importantly, there will be a precedent and a mechanism for setting parameters on domestic policy in an international forum. The U.S. Congress (and the European Council of Ministers) will have less control over U.S. policy, and will have to find less disruptive ways to achieve domestic policy goals.

Implications for U.S. Agriculture

U.S. agriculture has an important stake in seeing that trade continues to be open and free. Agriculture itself is highly protected around the world; many U.S. producers would benefit from more liberal agricultural trade. Moreover, U.S. agriculture depends on the less developed countries for its growth markets, and it is these countries who bear the brunt of new protectionism.

U.S. policymakers must continue to make the case for more open trade both in agriculture and manufactured products. There have been several efforts in the past few years to close U.S. agriculture off from the rest of the world. These attempts must be resisted. Policymakers must also be willing to do the hard thing—to make policy for the long, and not the short run.

Policymakers must make sure that U.S. agricultural policy does not restrict trade, or does not damage potential customers. The marketing loans on rice, for example, did a great deal of damage to Thailand, one of the United States important allies and trading partners. Such policies, however politically attractive, must be resisted if U.S. agriculture's long run prosperity is to be ensured.

Finally, U.S. policymakers must be willing to make changes in U.S. policies under the auspices of the GATT. Thus far, our actions—increasing planted acreage and continuing the Export Enhancement Program—speak louder than words. There is much bluster on Capitol Hill about setting agricultural policy in Washington, not Geneva. But U.S. agricultural policy options are already determined, in part, by policies set in Tokyo, Brussels, Canberra and Ottowa. Resisting an international agreement on acceptable agricultural policies will only serve to intensify the protection of agriculture overseas, and further restrict agricultural markets.

GLOBAL GROWTH

The Past Thirty Years

Between 1960 and 1980, global income growth rose steadily. For developed countries, GNP growth averaged over 4% during those two decades. Developing country incomes rose an average of 6%, and centrally planned economies grew at an average rate of 4%. Since 1980, the rate of growth in incomes has slowed. Developed country growth has averaged 2.5% annually; developing countries' income growth has dropped to 2.4%; and centrally planned economies growth has dropped to 3% (Economic Report of the President, 1988).

In some regions, per capita growth has actually fallen. In 1980, the average per capita income in Sub-Saharan Africa was $560; in 1987 it was $450. Per capita growth has also fallen in much of Latin America (Crook, 1989).

In part, some of the slowdown in economic growth is due to natural phenomena—income growth in the more highly developed countries will moderate over time—but some is related to thorny problems confronting the global economy. These problems include the debt crisis and misguided economic policies in both developed and developing countries. Yet, there are hopeful signs on the horizon as developing and centrally planned countries begin to recognize the benefits of market based economies.

Burgeoning Debt. The problem of international debt began with the first oil price increase in 1973. The OPEC countries were unable to absorb their new found wealth, and deposited their oil export earnings—which became known as petrodollars—into the Eurocurrency markets. From there, international banks lent, or recycled, the petrodollars to less developed and centrally planned countries in Latin America and the Eastern Bloc. At the same time, the developed countries, led by the

United States and Western Europe, tried to cushion the impact of higher oil prices on their economies by lowering interest rates.

With easy credit and favorable terms, borrowing by developing countries soared. Total external LDC debt rose from $70 billion in 1970 to $666 billion in 1981. Private, unguaranteed debt rose from $17 to $99 billion over the same period (World Bank, 1987b). The United States was a major beneficiary of this borrowing, which fueled an increase in exports to the developing countries. The share of U.S. agricultural exports to the developing countries rose from 10% to 43% between 1972 and 1981 (USDA, 1987a).

In 1979, the conditions that made borrowing cheap and easy reversed course. The developed countries responded to the second oil price increase by raising interest rates to offset the inflationary impact of higher oil prices. At the same time, the value of the dollar—in which most loans were denominated—jumped. High interest rates and expensive dollars combined to raise debt service payments in a time of global recession and shrinking global markets. The effect on the U.S. balance of trade was significant, falling over $10 billion with Latin America alone in between 1982 and 1985 (Economic Report of the President, 1987). Agricultural exports to the fifteen most indebted countries dropped from $39 billion in 1981 to $26 billion in 1986 (Food and Fiber Letter, 1987).

In the years since the debt crisis erupted, little progress has been made in solving the problem. In fact, many would argue that the problem has worsened. Between 1983 and 1988, economic growth in the most heavily indebted countries has averaged only 1.8% annually, far below population increases (Deane, 1989). Per capita income growth has shrunk by 1.5% in the highly indebted countries (Crook, 1989). One economist has estimated that if the growth rates of the 1970s had continued into the 1980s, the total income of highly indebted countries would have been $2 trillion higher (Tucker).

The response to the debt crisis can be divided into three phases. The first, crisis management, consisted primarily of stretching out the repayment schedule and imposing macroeconomic reforms in the hopes that the debtors' economies would recover their ability to manage their debt burden. The second phase began with the introduction of the Baker Plan, which attempted to encourage international banks to maintain their lending to the "obedient" debtor countries while simultaneously rescheduling their debts. The third phase began this year, with the Brady plan, which for the first time recognizes that some of the debt will never be repaid, and begins to make provisions for debt forgiveness.

While the Brady Plan has acknowledged the obvious, it also has highlighted the legacy of problems still confronting the highly indebted countries. First, debtor countries will find attracting additional capital extremely difficult after banks have been forced to write down their loans. Second, austerity measures, and the dearth of additional financing, have choked off much needed investment which could have helped countries regain their economic base. Third, it will be easier for the international financial system to ignore the plight of these countries once their debt is taken off the books of Chase Manhattan and Citibank.

U.S. Macroeconomic Policy. In 1980, Paul Volcker set out to rid the U.S. economy of inflation by sharply restricting monetary growth. At the same time, Ronald Reagan and Congress enacted a broadbased tax cut, and increased government spending. The first action reduced the supply of dollars; the second increased demand. The net result: a budget deficit of almost $200 billion and historically high interest rates.

In time, higher real interest rates, coupled with the relative strength of the U.S. economy, encouraged foreigners to invest in the United States by buying dollars. These foreign purchases drove up the value of the dollar. As the dollar became more expensive relative to foreign currencies, and as the U.S. economy grew faster than other developed countries, U.S. consumers began to increase their imports of foreign goods, while foreigners decreased their purchases of U.S. products. The net result: The United States became a net debtor, and is running a current account deficit of almost $200 billion.

These policies created problems for our international trading partners. The United States' inability to control its budget deficit kept dollar interest rates at historically, and comparatively, high levels. In order to prevent their currencies from losing strength to the dollar, Europe and Japan were forced to keep their interest rates higher than they might have otherwise chosen to. Moreover, the United States budget deficit has been absorbing foreign capital that might have been put to more productive advantage at home.

The United States has suggested that the solution would be for West Germany to stimulate domestic demand and Japan to open its markets. West Germany is reluctant to do so, for fear of fueling inflation. Japan has found it extremely difficult to open markets with any speed.

These policies have also placed U.S. macroeconomic policy on the horns of a dilemma. To stem the trade deficit, the United States must allow the dollar's value (and thus interest rates) to decline vis-a-vis our trading partners. But, to continue to attract financing for the budget deficit, the dollar's value (and interest rates) must remain high in relation to other currencies, and the U.S. economy must remain strong. In addition, to control inflation, interest rates must be maintained; but, to avert recession they must not be kept up too high. Moreover, very high interest rates risk sending heavily indebted countries, companies and consumers into bankruptcy. Finally, in the event of a recession, the traditional macroeconomic solution—to expand the budget deficit—may not work, and may in fact be harmful if it leads foreigners to dump dollars.

Steering a course between high, but not too high interest rates will be extremely difficult and risky. A flight from the dollar could prove devastating—and is not unimaginable given the size and strength of the world's capital markets.

Developing Country Policies. When economists began to turn their attention to the challenges facing Africa, Asia and Latin America in the 1950's, the theory of economic development was in its infancy. Most economists argued that the developed countries' demand for primary products produced by the less developed countries would decline over time. The prices of these products would also fall over time, leaving developing countries with declining export revenues. Clearly, these economists believed trade could not serve as an engine of growth for the less developed countries.

These early development economists prescribed policies of import substitution, where poor countries were encouraged to import machinery and various intermediate goods to build an economy that could compete with wealthy, industrialized countries. These policies were a disaster. Not only did they leave developmental dinosaurs—vast infrastructures and industries for which there was no demand—they also squelched exports and did not succeed in controlling imports. The most important problem, however, facing the less developed countries was managing this economic system. Interventions piled upon intervention, and the private sector was pushed underground, or squashed altogether.

Some countries did not follow this inward looking advice. Instead, for a variety of reasons, they promoted exports. The economic results are startling. Countries with strongly outwardly oriented policies grew by an average of 8.6% annually between 1963 and 1985. Countries with a strongly inward looking strategy grew at an average annual rate of 3.3% over the same 22 year period (Crook, 1989).

Despite these lessons, many countries continue to follow inward looking strategies, with burdensome bureaucracies. Among those countries counted as inward looking are the highly indebted countries, and much of Africa. These misguided policies facilitated the accumulation of debt and have led to declining per capita growth rates—and much human misery—around the world. They have also put a drag on global growth, and on trade growth.

The Shift to Liberalization. In the past decade, the wisdom of following an outwardly and market oriented development strategy is beginning to prevail. The success of the Asian tigers (Hong Kong, South Korea, Taiwan and Singapore) and of China has attracted much imitation. Countries from Bolivia to Poland to the USSR are in the midst of liberalizing their economic (and political) life. The results in Africa have been astounding, particularly in agriculture. In countries with strong reform programs, agricultural production in the last three years has doubled since 1980-84. Agricultural production in reformist countries is also double that of countries that have not pursued reforms. Similar statistics can be cited for exports and domestic investment.

This movement towards liberalization is by no means widespread. It is however an important change in economic policies that will determine the world's future capacity to grow.

The Next Thirty Years

For the time being, the debt crisis has abated. The debtor countries are struggling, but the international financial system appears to have muddled through. However, the crisis could erupt again, if the global economy faced another recession, or if oil prices climbed or if commodity prices plummeted. Unbridled protectionism could also reignite the crisis. As long as the debt problem remains unresolved, global growth and trade will lag far behind their potential. The debt will continue to be a major problem facing policymakers in the next thirty years.

The imbalance of U.S. macroeconomic policies is the second major problem confronting policymakers. Despite promises and automatic sequestering, the United States has failed to bring the budget deficit under control. Again, no crisis has erupted. Thus far, foreigners seem willing to continue buying dollars to finance U.S. consumption. However, crisis lurks just below the surface. A sudden switch away from dollars or a recession could unleash a catastrophe.

For the last thirty years, development policies have been misguided. Increasingly, less developed countries are recognizing the need for more outward-looking, market-oriented policies. Shifting from existing policies will impose great costs in the short run, however, that will need to be borne in part by developed countries and development institutions. Clearly, this shift from stifling to liberating policies should be encouraged by all who have an interest in trade.

In sum, achieving strong global growth will require careful management and wise decisions. The economy is balanced on a knife's edge. The room for maneuver is not large, and mistakes either way could have tragic implications for the world economy. Economic cooperation will be critical to the success of management, but cooperation is never easy when the challenge is problem prevention.

Implications for Agriculture

Clearly, U.S. agriculture wins when the world economy and world trade are growing. It is especially important for U.S. agriculture that the less developed economies succeed, because there are the largest potential growth markets. The problems confronting the world economy—debt and deficits—are beyond the scope of agricultural policy, yet, agriculture can contribute to their solution. First, attempts to subvert development assistance to potential LDC competitors (such as to soybean producers in Brazil, or wheat producers in Argentina) are short sighted and counterproductive. U.S. agriculture's longer term interests will be best served by having strong economies in the less developed world, who in time will become strong markets.

Second, U.S. agriculture can contribute to reducing the pressure on the budget. While it is foolish to ask agriculture to forgo its share of the pie if no one else is offering to forgo theirs, it is equally foolish for agriculture to continue to insist that it should somehow be exempt from the budget cuts that must fall on everyone.

Third, U.S. agriculture can support liberalization efforts in less developed and centrally planned economies by opening its protected markets in sugar and dairy, and reducing support in other commodities under the auspices of the GATT. Only through example and with available markets will liberalization efforts in these countries succeed.

SUMMARY AND CONCLUSIONS

The next thirty years will see a continuing decline in the relative strength of the United States. This decline should not be seen as a drawback, for it also means that other economies are becoming healthier and stronger, and that the fruits of the world economy are being shared more widely. The declining relative importance of the United States in the world economy will present challenges. Cooperation with other players will become more important and more necessary. The United States may continue to exert political leadership, but it will need the support of a number of other governments if it wishes to exert economic leadership.

The less developed countries hold the key to the global economy's future. While strong growth in the developed countries will probably insure against global recession, without growth in the less developed countries, the scope for world trade will be sharply limited. The debt crisis and policy reform will be the major challenges facing policymakers. Economic liberalization will prevail in the developing and centrally planned economies. While liberalization will not be easy, it will be pursued. The Bretton Woods institutions (the GATT, the IMF and the World Bank) can and should all support this process. Ironically, demands for protectionism will continue to increase in the so-called liberal democracies as increased competition for newly industrializing countries threatens employment in some economic sectors. Again, the Bretton Woods institutions can play a useful role in dampening these demands. International macroeconomic phenomenon will continue to dominate U.S. domestic policies, and U.S. domestic policies will be increasingly affected by other countries' domestic policies. The only solution will be policy reforms that allow for macroeconomic fluctuations and that take account of others' policies.

There are several trends outside of international economics that are worth noting. First, is the death knell of the Cold War. It may take fifty years for the economies of the Soviet Union and the Eastern Bloc to become fully integrated into the global economy. But that process is clearly starting, and is likely to be irreversible. Second, is the continued pressure of population increases in the less developed countries. While the world is physically capable of feeding these new mouths, it will require sustained effort on the part of agricultural and development policymakers to ensure that they do indeed have access to food. Third, the global economy of the last thirty years has been dominated by the triumvirate of Bretton Woods institutions. The changes in the world economy that have occurred in the last decade have placed all three institutions under strain. A critical test of the next thirty years will be to adapt these institutions to meet the challenges of a multipolar, integrated economy.

THE IMPLICATIONS
FOR AGRICULTURAL CURRICULUM

If I were designing a curriculum for agriculturalists graduating in 2020, I would include coursework in macroeconomics, development economics, and international trade. I would insist that the focus of these economics classes not only be theoretical, but that a sound basis in policy prevail. The challenges facing agriculture in the next three decades will need a broad understanding of the way the world economy functions, and how it relates to agricultural problems. The challenge will also be to understand the role of policy in confronting these macroeconomic challenges and to design new policies that better take account of the new realities facing the agricultural sector.

Finally, I would require students to become more attuned to other cultures. The economy is global, agribusiness is global, agricultural markets are global. Agriculturalists who do not understand languages, and do not appreciate cultural differences in tastes and behavior will be ill-equipped to face this global environment.

These are fairly straightforward and simple ideas to confront what seems to be a complex set of challenges. But, what I am suggesting is that the current agricultural curricula are often too narrow in scope. It teaches students a great deal about the wheat and corn markets, but not enough about the Eurocurrency markets. It concentrates on the ins and outs of U.S. commodity policy, without sufficient attention to U.S. macroeconomic policy. Broadening the students' horizons in this manner may conflict with the needs of Ph.D.s to "know more and more about less and less"; but given where the world is heading, educating generalists may ensure that U.S. agriculture continues to meet the challenges ahead.

NOTES

The author would like to thank George E. Rossmiller for his insights and comments. This paper was prepared for the University of Wisconsin's Strategic Planning for Agriculture and Natural Resources Steering Committee. It was presented on November 7, 1989 in Madison, Wisconsin.

BIBLIOGRAPHY

Akers, John F. 1989. "What It Takes to Compete." *Financier* July, 21.

Artus, J. R., and J. H. Young. 1980. "Fixed and Flexible Exchange Rates: A Renewal of the Debate." Washington, D.C.: International Monetary Fund.

Baldwin, Robert E. 1981. "The Tokyo Round of Multilateral Trade Negotiations" in Robert E. Baldwin and J. David Richardson, *International Trade and Finance*. Boston: Little, Brown and Company.

Berry, John. 1989. "If a Dollar Crisis Hits, Europe Might Not Mind." *Financier* January, 5.

Blandford, de Gorter, Dixit and Magiera. 1988. "Agricultural Trade Liberalization: The Multilateral Stake in Policy Reform." unpublished paper.

Borster, Tim. 1989. "Latin Debt Crisis Still Shifting Trade." *Farmline* August, 4–6.

Crook, Clive. 1989. "The Third World Survey." *The Economist* September, 23.

Corden, Max. 1984. "The Rival of Protectionism." New York: Group of Thirty.

Deane, Marjorie. 1989. "Debt Arguments Will Dominate IMF—World Bank Meeting." *Financier* September.

———. 1988. "At Quiet Bank-Fund Meeting: Thoughts of Monetary Reform." *Financier* November.

Economic Report of the President, 1988, 1987, and 1986.

Emmett, Bill. 1989. "For Japan, Not World Dominance, But Leadership of A Year Bloc." *Financier* January, 2:1.

European Agricultural Policy Issues and Options to 2000: FAO 5/8 ERC/88/INF.

FAO. 1986. World Food Report 1986 (Roma, Italy).

Farmline. 1989. "Austerity Policies No Solution for Brazil." August, 6–8.

FATUS. 1989.

Feldstein, M. 1987. "The End of Policy Coordination," *Wall Street Journal* November 9.

Financier 1988. Editorial. February, 12:3.

Financier 1989. "Interview with Maurice Strong: Adaptations of the Bloc's." April.

Food and Fiber Letter. 1987.

Garten, J. E. 1985. "Gunboat Economics." *Foreign Affairs*. February.

Group of Thirty, 1984.

Hennes, John M. 1989. "Risks, Rewards of Major Trades in Global Deregulation." *Financier* August.

International Monetary Fund, 1987. International Financial Statistics 1987 Year Book (Washington, D.C.).

Karlik, John R. 1981. "Some Questions and Brief Answers About the Eurodollar Market" in Baldwin, Robert E. and J. David Richardson *International Trade and Finance*. Boston: Little Brown and Company.

Kost, William E. and Cathy L. Jabara. 1985. "Agricultural Trade is Vital." Agriculture in a Global Economy. *1985 Yearbook of Agriculture*. Washington, D.C.: USDA.

Longmire and Morey, 1983. "Strong Dollar Dampens Demand for U.S. Farm Exports." USDA. ERS, Foreign Agricultural Economic Report, no. 193.

Marris, Stephen. 1987. *Deficits and the Dollar* Washington, D.C.: Institute for International Economics.

McCalla, Alex F. 1989. "Emerging Patterns of World Agricultural Trade." Paper presented to Plenary Session of the Annual Conference of the Agricultural Institute of Canada July 10.

McCracken, P. 1987. "Toward World Economic Disintegration," *Wall Street Journal* February, 9.

OECD, 1983. Committee for Agriculture. Issues and Challenges for OECD Agriculture in the 1980s. April. AGR (83) 15.

Ostry S. 1987. "Amid Complex Global Linkages, Opportunity and Vulnerability," *Financier* November, 11:11.

Paulino, Leonardo A. and Mellon, John W. 1984. "The Food Situation in Developing Countries: Two Decades of Review." *Food Policy* November, 9:4.

Rappeley, W. 1989. "A New Architecture," *Financier*.

Roosa, Robert V. 1982. "The Multiple Reserve Currency System," *Reserve Currencies in Transition*. New York: The Group of Thirty.

Rossmiller. 1986. "Farm Exports: An Historical Perspective," *Choices* Third Quarter.

Sanderson, Fred and Rekha, Mehra. 1988. "Brighter Prospects for Agricultural Trade" in M. Ann Tutwiler, ed. *U.S. Agriculture in a Global Setting: An Agenda for the Future*. Washington: Resources for the Future.

Sease, Donald. 1989. "Program Trade Stirs Stock Surge," *Wall Street Journal* September, 29 C1

Shane, M. 1987. *Agricultural Outlook* August, 18–21.

Sullivan, Barry. 1989. Interview. *Financier* August, 13:18.

The Economist 1989. "Tough Traders Come Last." September, 9 17.

Tucker, S. "The Legacy of Debt" Overseas Development Council, Policy Focus 3.

Tutwiler, M. Ann and Barbara J. Elliott, "An Interdependent and Fragile Global Economy" in M. Ann Tutwiler, ed. U.S. Agriculture in a Global Setting: An Agenda for the Future, Resources for the Future. Washington, 1988.

Uchitelle, L. 1988. "Two Hard-to-Quit Habits Sustain Trade Deficit," *New York Times* January, 14 D1.

"U.S. Trade Policy and the GATT: Implications for Agriculture." Unpublished paper. David Blandford for NCFAP, Assessing the Implications of Alternative Mean of Supporting Agriculture, October, 1989.

United States Department of Agriculture. 1983. Economic Indicators of the Farm Sector: Income and Balance Sheet Statistics. Washington, D.C.: Economic Research Service.

_____. 1986. Economic Indicators of the Farm Sector: Income and Balance Sheet Statistics. Washington, D.C.: Economic Research Service.

_____. 1987a. Agriculture Yesterday & Today, Washington, D.C.

_____. 1988. World Agriculture Situation and Outlook, Economic Research Service WAS53, December.

World Bank. 1987. Developing Country Debt, Implementing the Consensus. Washington, D.C., February.

_____. 1987a. World Development Report 1987, Washington, D.C.

_____. 1987b. Developing Country Debt, Implementing the Consensus. Washington, D.C., February.

5

For 2020: Trends and Issues in Agricultural Production and Marketing

DUANE ACKER

From the myriad issues that will certainly face U.S. agriculture and natural resources in the early part of the twenty-first century, a few, I believe, will be paramount. Several of these are currently receiving far too little attention.

Our task in colleges of agriculture and natural resources is to serve human beings over time and to serve them well. Humans seek nutritious food that will enhance physical and mental well-being, productivity, and longevity. They want food that is appealing and secure in quality; blue sky, clear and clean water and air; green grass, flowers, shrubs, and other features of an attractive landscape. Having worked in an agricultural area in the Midwest before moving to Washington, I realized the importance of amenities like clear water and air. During the time we were in South Dakota, we had a Congressman who came out from Washington to see what a watershed looked like. He said that for 22 years he had voted for watersheds but had never seen one. The Congressman had never been on the ground west of Chicago. We met him at the Sioux Falls airport and took him outside into a cold, crisp, clear night. He took a deep breath of that clean, fresh, South Dakota air and collapsed. We had to pull him over to the curb and put his head under the exhaust of a bus.

I will discuss ten important issues facing U.S. agriculture and our natural resource base as we enter a new century. Since people of the world are consumers, I direct your attention first to the general area of marketing.

ECONOMIC GROWTH
IN LESS DEVELOPED COUNTRIES

In recent years, U.S. agriculture, has exported 40 to 60% of some of its major farm commodities. The U.S. has unused production potential and would like to export more. However, there is not much market growth potential left in the U.S., Western Europe, or Japan.

Except for the residents of less developed countries (LDCs), who would like to eat more, the sector of the world most likely to gain from increasing income levels of the world's 3.5 billion people is U.S. agriculture. Where family income is $200 to $600 per year, about 60 cents of every dollar increase in income will be spent on food—either more food or food of higher nutritional quality. In addition, low income countries have the highest rates of population growth.

What will help LDC economies grow? In many cases, developing the agricultural sector, that currently employs 50 to 85% of the labor force, is the first step. Yet today, some in U.S. agriculture see LDCs only as competitors. Producers and students in colleges of agriculture and natural resources must develop an understanding of market potential. Some of my neighbors in western Iowa are a little suspicious of the process of increasing market potential in the Third World. Indeed, many people look at it that way. However, an example will illustrate the process and the outcome. In 1954, Korea was an economic basket case. The U.S. and some other countries developed short-straw rice varieties that used water and fertilizer more efficiently and gave higher yields. They also helped to develop an extension service and agricultural experiment stations. It took a long time, but productivity gradually increased. By the early 1970's, the U.S. was sending Korea annually about 200 million dollars worth of food aid, but Korea was buying another 200 million dollars worth of agricultural products from the U.S. In the early 1980's, the U.S. no longer sent food aid to Korea, but Korea bought an average of 1.8 billion dollars worth of U.S. agriculture commodities per year. During the ten year period from the early 1970's to the early 1980's, per capital production of agricultural commodities in Korea increased by 27%.

Colleges of agriculture teach freshmen in animal science that the stomach does not expand much. In the U.S., most people are eating as much food as they want. However, if people are not eating all they want and their income rises, they will eat more. Not only more food—a shift from 1500 to 2500 calories per day, for example, but different foods—a shift from rice to meat—first chicken, and then maybe beef or pork, depending on the culture. What does it take to produce chicken? It takes corn and soybean meal. Who sells corn and soybean meal? The cycle benefits not only U.S. agribusinesses dealing in these commodities, but the fertilizer supply business, the machinery business, the seed business, and the food processing business. They each hire people and the cycle begins to get larger; the economic machine begins to roll.

MEETING CUSTOMER NEEDS
AND PREFERENCES

Retailers and brokers in Hong Kong and Singapore tell me that the U.S. could sell many more U.S. grocery products in their countries if U.S. firms: 1) use packages and labels that hold up in hot and humid conditions, 2) provide what consumers want, 3) offer to ship less than a container load of a product retailers want to try, 4) send only their most effective salespeople, and 5) at a minimum, use liters and grams instead of only quarts and ounces on the labels. There is a simple message used by an insurance agent to train his new employees that seems applicable to the food industry. The agent said, "People buy not because they are made to understand, but because they feel they are understood."

In other words, U.S. businesses need to pay attention to cultural practices around the world. For example, who eats pork and who doesn't? Who prefers lamb or goat over beef? Which cultures prefer beef? Who wants white corn or sorghum for laying hen rations so egg yolks will be light in color? Who wants soft wheat, hard red winter, feed wheat? The U.S. lumber industry has had difficulty selling plywood to Japan; the Japanese do not use 4 feet by 8 feet sheets.

Understanding preferences of consumers from around the world begins with the curriculum and degree requirements established by colleges of agriculture and natural resources. How many college of agriculture graduates know a second language? How many know the culture or some of the customer preferences in China, India, Nigeria, or Indonesia— four of the most populous countries of the world? Or of Japan, South Africa, Germany, France, Canada, or England—some of the countries with the highest average income levels in of the world?

The U.S. has been self-sufficient to its very severe detriment. In fact, economic self-sufficiency may have been more costly to the U.S. from a social and cultural standpoint than the Great Wall has been to China over the decades. For the U.S., there is a tremendous cost associated with the rest of the world adapting the English language. I come back to the curricular issue and raise the question, "Why is it that when faculty talk liberal arts and the general education of our students we always talk about western civilization as being so important?" Most of our ancestors came from Western Europe. But, how about all the folk who moved from Mexico and Central America into New Mexico and Colorado, then spread across Texas, Oklahoma, and Kansas? How about the Southeast Asians who came to California and whose children and grandchildren are our engineering school colleagues? Should university curricula not give equal treatment to the cultures of Central America and Southeast Asia?

GLOBAL ECONOMIC INTEGRATION

Made possible by both transportation and electronics, global economic integration has allowed major economic change anywhere in the world to have some impact on almost any agricultural input supplier, producer, processor, or marketer. Let me emphasize particularly the impact of international capital markets on U.S. agriculture, because international money movement dwarfs international commodity trade.

In recent years, the volume of U.S. agricultural commodities moved worldwide increased astronomically. The increase was about sixfold—from 10 to 14% of the total volume of agricultural commodities exported, to 40, and then 60% marketed outside the U.S. In contrast, in 1984 money movement was 10 times the dollar value of world commodity trade. During 1989, worldwide currency trading totalled more than $300 billion per day. Instantaneous transactions that change relative values of currency may be more important to a business that deals internationally than the production cost of a commodity or its position. When you examine the sharp drop in U.S. agricultural exports in the early 1980s, understanding capital markets is critical as is understanding agricultural policies, like the floor put under prices in the 1981 Farm Bill.

Together with this monumental change, consider a few other major economic changes that illustrate global integration:

1. the European Economic Community and its movement toward free trade internally, a single currency, and a consolidated economy by 1992;
2. the economic growth of Southeast Asia—Indonesia, Singapore, Malaysia, Thailand, Hong Kong, Taiwan, and others;
3. the economic growth and prominence of Japan;
4. the heavy debt load of the U.S.—moving toward $3 trillion;
5. the heavy debt load of many developing countries;
6. movements toward a market economy in China, the Soviet Union, Poland, and several other countries; and
7. movement toward debureaucratizing within some highly bureau-cratized market economies, like Chile, that allow easier importing, exporting, or accepting of foreign investments.

These changes are the result of, and also prompt, more rapid worldwide communication; more efficient handling of cargo (containerization and roll on/roll off); round-the-clock trading of currencies, equities, and commodities; and greater attention by GATT (General Agreement on Trade and Tariffs) negotiators to reductions in both tariff and non-tariff barriers.

In the U.S., we used to have "our own territory." Today, the world is our oyster, and that of everybody else.

TRADE RESTRICTIONS

The U.S. posture regarding free and open world trade of agricultural commodities and products is rather clear. Some suggest that there is not as much unanimity on that posture as appears, but would-be objectors within U.S. industry do not believe it will be achieved in GATT negotiations anyway. Regardless, steps toward free trade are likely to continue. U.S. agricultural people need to understand both the potential benefits and the risks of unrestricted trade.

Agricultural producers, processors, wholesalers, and retail firms need to understand what U.S. commodities will suffer from more free trade, and in which sections of the U.S. they may suffer most. We also need to understand secondary and tertiary consequences of more free trade. In this way, people can plan long-term adjustments.

We usually limit our thoughts to "Can we export more?" or "Will we be hurt by competing imports?" But other countries trade with each other; we are in a global market so there are several additional questions. Will more free trade enhance world economic efficiency? Will it increase total world consumption? Will it enhance our comparative advantage?

What about less developed countries? Many of the LDCs have been given special dispensation by the U.S. and other developed countries on import tariffs, and many maintain restrictions on imports to their countries to "protect a developing agriculture" or to save foreign exchange. However, a good many studies suggest import restrictions actually slow economic progress in less developed countries because they inhibit trade among the LDCs, themselves. Trade among the LDCs can be very significant and can help each country find their comparative niche. As the Korean example indicates, of paramount importance to U.S. agriculture is that LDCs achieve economic growth as rapidly as possible.

As an industry, we should not worry about our ability to shift enterprises. During our lifetime, major centers of cattle feeding have moved from the Corn Belt to the High Plains, egg production from the northern Midwest to the Southeast (and now there is some movement back to the Midwest), cotton production from the Southeast to the

Southwest, and feeder cattle production to land in the Southeast formerly used for cotton.

GLOBAL COMPETITION: COST PER UNIT
OF QUALITY AT THE DESTINATION

While cost per unit of quality applies as well in the domestic market, I emphasize the world market because U.S. agriculture depends heavily on the export market. We have a high proportion of Class I land, a temperate climate, and a good reputation for volume and quality. But other countries have considerable production potential, generally lower labor costs, may be closer to some of the consumers we would like to serve, and also have good reputations for quality.

A major concern, therefore, is whether we can afford to continue some current practices and remain competitive in the new global economy. For example, can we afford to continue shipping up to three per cent foreign material, and sometimes even more, in our exported grain? Or should we keep the grain clean after it leaves the combine and clean any that is delivered to the wholesaler with foreign materials, in order to save the cost of shipping the foreign material and the cost of the miller removing it? Someone will discover how to eliminate the dis-economy of foreign material in grain and make a profit on it. It might as well be my student as someone else.

We buy and sell fertilizer on price per unit of nitrogen, milk on the basis of milk solids, and finished livestock on carcass weight and grade; perhaps we should negotiate grain sales on price per unit of clean grain. The same principles hold for cotton, sugar, meat, and especially processed products.

Lastly, I ask, "Can we afford to continue to require that a percentage of exported grain (PL 480) be shipped on U.S. vessels, often at a higher cost than on other vessels? Can we afford to continue to require that a percentage be shipped from Great Lakes ports?"

FOOD QUALITY

The European Economic Community (EEC) wants hormone-free beef. Australia lost the Singapore egg market to "hi-tech" egg production units on the island; eggs with the "hi-tech" logo are less than a day from the hen when they arrive at the display case. In the U.S., the demand for oats skyrocketed when oats was linked to lower cholesterol levels. In recent years, as a relationship between calcium intake and prevention of osteoporosis has been debated, producers of dozens of products advertise "calcium added."

I talked with my daughters—one a physical therapist and the other an attorney—about the scientific evidence surrounding the need for hormone-free beef. They said, "Well, who cares about scientific evidence. What's wrong with selling hormone-free beef? Maybe that is what we will buy; we have fairly good incomes." The point is that though the U.S. consumer may be on a "health kick" that will diminish, and though recent media hype (both justified and unjustified) regarding food risks will likely ameliorate, there is no question that a heightened awareness and concern among consumers about food quality will endure. Most scientific studies report simple correlations, such as between fat intake and blood cholesterol level. Forms of fat, forms of cholesterol, effects on other body measurements, and interactions with carbohydrate intake or life style have been looked at in only a few cases, but more studies with interpretation by knowledgeable and prudent scientists are essential.

In the meantime, should U.S. producers fight bans, such as the one on hormone-fed beef by the EEC as "non-scientific," or meet the market with beef certified as hormone-free? I am reminded of a story told by a former extension agent about a buddy who had a reputation for being a tremendous fisherman; he always caught fish even on occasions when others did not. His friends were a little suspicious, so they arranged for their friend to take a local conservation officer on a fishing trip to show him how to fish. A half hour into the fishing trip, the two men had caught no fish and had had few bites. The buddy reached under the seat of the boat and pulled out a stick of dynamite, lit it, and tossed it into the water. The fish floated to the top. The conservation officer said, "Man, you're in deep trouble" and he began to lecture on the morality of fishing. The buddy took out another stick of dynamite, lit it, and handed the dynamite to the conservation officer, who took it unknowingly and kept talking while he pulled out his badge. The buddy said, "Are you going to keep on ranting, or are you going to fish?"

The application of the story is to meet consumer desires. Our producers believe in free enterprise; they believe in competition. They believe in the principle of finding a marketing niche and serving it. Apply these concepts to the hormone-free beef issue. Likewise, if people want pork from pigs that are raised unfeathered, should they not have that privilege? Some entrepreneur will find these market niches. Is there any reason why it should not be one of our students? From one of our extension clients? From a producer in Wisconsin rather than an entrepreneur from Salina, Kansas? While negotiating vigorously on artificial trade constraints at GATT, I would be helping my clients find alternatives.

Finally, technology allows dating and tracing of food items, both ingredients and finished products. It also allows precise measurement of microcontent. Within the food industry, aggressive and alert producers, processors, and marketers will refine and use this technology to insure the quality of the food product and to assure consumers. The issue remains whether additional nutritional labeling should be required or be voluntarily done by industry groups.

CONTRACT PRODUCTION AND PROCESSING AND OTHER FORMS OF RISK REDUCTION

The trend toward contractual integration in an increasing number of commodities and products is clear. As size of enterprise increases, as consumers appear more fickle (or responsive to new information), as global competition increases, as the effect of global economic integration becomes more evident, as we depend more on borrowed capital, as interest rates and other costs remain high, and as managers at all points in the agricultural chain become more precise in their financial planning, each actor in the chain tries to reduce risk. A series of articles in the Des Moines Register comparing the Iowa and North Carolina swine industries illustrates the issue.

Iowa has long been the leading state in pork production, but North Carolina is growing rapidly, largely with contract production. North Carolina imports corn at a higher price per bushel than the price paid by the Iowa producer. But large units, standard management regimes, willingness to care for animals on a per-head basis, a limit to risk, and other factors have brought about rapid increases in swine numbers in North Carolina. There are other issues. Iowa farmers tend to be more independent, have had higher incomes, and a higher proportion have operated their own units. They are more reticent to follow someone else's prescribed management pattern. Yet, the proportion of Iowa swine being produced and marketed under some form of contractual arrangement appears to be increasing.

The impacts of contract production extend beyond numbers of animals produced. They include the effects on demand for grain and grain pricing; on feed manufacturers, dealers and salespeople; on demand for extension education programs and on college graduates?

Finally, contract production and processing trends raise questions about efficiency, quality, consumption, and risk reduction. Some of the most important are: What forms of integration are most efficient? What forms tend to put the highest priority on quality? Do they allow development of brand loyalty? Can they bring about increased consumption? Can they be more effective in achieving exports and competing on the world market? Are the commodity futures and options markets properly structured and operated? Do they allow input suppliers, producers, and processors the risk reduction they need? Are managers equipped to use commodity futures and option markets well, and in concert with other risk reduction strategies?

Producers use all forms of contracts to reduce our risk—we contract diesel fuel and fertilizer ahead, for example. Today, there are quite a few producers in the U.S. who use currency futures markets to protect themselves; it is so logical. If fluctuating currency values have more impact than the production cost of a commodity—and in some instances that is the case—then it seems prudent to protect oneself through these vehicles if one is producing an export commodity like soybean meal or corn.

THE NATURAL RESOURCE BASE
AND THE ENVIRONMENT

Though the term environment usually heads any discussion of this type, that term is so broad it sometimes prevents a focused discussion. I choose, instead, to address the soil, the ground and surface water, the plant and animal species, and mineral deposits, plus the air and aesthetic features of our surroundings.

I need not repeat all of the concerns expressed in recent years, nor catalog the effects that are evident in many communities (e.g., groundwater with too-high nitrate levels, eroded hill-sides, silt accumulations in major reservoirs). Issues such as these are a major concern of all Americans because they affect our ability to sustain a productive agriculture and a pleasant and healthful place to live.

I want to zero in on an environmental issue that I do not think has received enough attention—genetic diversity. The question is how much genetic diversity must we retain, or have access to, for our U.S. corn, wheat, oats, forage, or vegetable systems, in order to insure that a new disease will not cripple an industry segment for a sustained period? A narrowing of the genetic base of most of our major crops is of major concern when one has lived through a 10 to 15% loss of a crop that could easily have been 75%. Remember corn leaf blight of the early 1970s and the Texas male sterile genetic line that was almost universally used in seed corn production at that time?

As we continue genetic progress in Holstein dairy cattle, with mathematical precision and electronic speed, toward ever-higher production, and as we accentuate that progress with cloning—do we risk developing a too-narrow genetic base in our U.S. dairy system?

A final example: on the Great Plains spanning four states, grain sorghum is an important crop. The producers buy their seeds from perhaps 20 companies; probably 80% of the seed is from 10 companies or less. They all use in-breed lines from the experiment station that put them together. With this narrow of a genetic base, what would happen if an exotic disease develops? In contrast, look at India, a country that has between 3 and 30 million people raising grain sorghum. In India, they have been growing grain sorghum for 6 centuries or longer. They have high altitude and low altitude varieties. Where are you going to find the genetic diversity, in Kansas or India?

Let me close this section by drawing on my work with the U.S. Agency for International Development (USAID). There must be 200 natural resource and environmental organizations with headquarters in Washington, D.C. that are keeping an eye on the work of USAID and the World Bank. When I first went to Washington, one of the things I encountered was an effort by many of these organizations to bring about the creation of a separate environmental department in the World Bank and USAID. The objective of such a department was to protect the natural resource base in developing countries by scrutinizing every project funded by these agencies.

My approach in USAID was to elevate the environmental sensitivity of the soil scientist, the entomologist, and the animal scientists, by putting a natural resources specialist on the project team. First of all, the approach avoided duplication of effort of one unit checking on another that often creates conflict within the organization. Secondly, it lead to the creation of a dual statement on maintenance and enhancement of the natural resource base. After two years, one of the representatives of a major environmental groups went to Capitol Hill and recommended that the agricultural program of USAID be increased 50 million dollars. He liked what was being done by USAID in some of the developing countries.

TECHNOLOGY AND SOCIETY

Six supermarket chains recently announced they would not handle milk produced from cows receiving bovine somatotropin. Some countries have imposed constraints on caged pullet and layer operations. Will there be similar negative reactions to embryo transplanting, sexing embryos, or gene manipulation? Why has there not been a negative reaction to artificial insemination, meat tenderizers, or hydroponics?

Some of the negative reactions from consumer or other groups reflect concern for animal welfare or comfort; with others, the concern appears to be the apparent "exploitation" of animals. What explains these reactions? Is it the social environment or prevailing cultural themes at the time the new technology becomes well-known?

Will constraints on adoption of technology brought about by the unwillingness of consumers to accept the output diminish our world competitiveness? To what extent? At what cost? In what areas does the U.S. populace want to insure U.S. competitiveness? Is agriculture one of these?

LAND OWNERSHIP AND CONTROL

In some states, corporate ownership of farmland occupies an annual spot on the legislative agenda. In several, it emerges only when an outside corporation proposes to invest in a major enterprise in the state. More recently, land ownership by foreign citizens or corporations has emerged as a public issue. However, although less concern has been expressed, there has been much more foreign investment in processing and marketing enterprises. These industries are as critical or "controlling" in the food chain as producers.

As we move toward a fully integrated global society, will we not see more inter-country investment? If we seek global peace and trade, aren't such arrangements contributory?

In Central and much of South America, a major concern is for the landless, the family that would like to own or control a piece of land but cannot. That hasn't been much of an issue here, but it could become one. Should our immigration rates accelerate, or should there eventually develop some form of "common market" and free people movement among Canada, the U. S., and Mexico, the scene could change drastically.

IN CONCLUSION: TWO SUGGESTIONS

I realize that my major role has been to identify issues rather than to list solutions or actions. But, I cannot resist offering two suggestions. The first is to underscore that the thinking process guiding curricular, extension, and research endeavors should quite often block out "the current way of doing things" or the "current way of looking at things." I can hear the response to my comment about shipping dirty grain, "Our clients—exporters, elevator operators, and others—would send clean grain if they would get paid for it." In preparing our extension clientele or students for the future, and in designing our research, that type of response cannot invade the thinking process. The world economy will not tolerate forever the diseconomy of adding, shipping, and removing low-value, foreign material. Some creative or determined person will find a way to eliminate that diseconomy, to his or her considerable profit. It could be one of our students or one of our clients.

The thinking process of the dean, the department head, the curriculum or program committee, and especially, the individual faculty member, must be on the core issues: where market growth *can* occur; food quality attributes (rather than on weight or volume); the U.S. economy as an integral segment of the global economy; risk reduction; the integrity of the water and soil; and the essence of technological consequences (including the respiration or pulse rate of caged hens or the comparative attributes of milk from somatotropin cows).

I repeat, sometimes we need to block out the old ideas. We might be like the Australian who got a new boomerang and tried to throw the old one away. He had a heck of time getting rid of it; he just could not throw it hard enough.

My second suggestion is to develop a mission statement for colleges of agriculture and natural resources. People want to know what is the college's purpose for existence. There is some concern that the agricultural sector behaves defensively too quickly and too frequently. USAID programs were on the defensive vis-a-vis U.S. farmers who were suspicious that USAID was building competition. So we created a statement that read:

> The goal of the food and agriculture programs of USAID is to increase the income of the poor majority, and expand the availability and consumption of food, while maintaining and enhancing the natural resource base.

We emphasized this approach to every staff member, every mission director, every project leader, every contractor, as well as every member of Congress on the Appropriations Committee.

Perhaps the colleges of agriculture and natural resources, the agricultural industry, major cooperatives, and input suppliers each need a mission statement. The statement could take the form of a Hippocratic Oath for faculty that says:

> I pledge to discover and teach the biological, physical, economic, and social facts, concepts, and principles that will allow (Wisconsin) agriculture to consistently, and with reasonable expectations of profit, provide its clientele adequate quantities of nutritious, wholesome food that will enhance both physical and mental status and productivity over the long term; to nurture plant and animal life that will persist and that will enhance the aesthetics and pleasures of life, making Wisconsin a nicer place in which to live and work; and to promote the political and social structures that will most benefit human interaction, while maintaining and enhancing the

natural resource base and demonstrating sensitivity to the biological and social comfort of living things.

The above is far too long and no doubt either too narrow or too inclusive for some. But every entity needs to remind itself, and its clientele, in some succinct and expressive manner, of its greater purpose for existence.

6

Forestry and the Environment: From Spare Time to Big Time

JOHN C. GORDON

INTRODUCTION

Research and education must discern where society is going and be there ahead, to meet it. They can do this through leadership or prescience or some combination. Environmental issues, broadly construed, are now in the public eye, and forests and forestry are explicitly included. But the call is for solutions, not more horror stories. Politicians seem willing to help. What will universities do? Will they choose an effective combination of leadership and vision? Or will they miss the meeting?

Three related societal themes underlay the resurgence of environmental concern and debate: sustainability, global change, and human population. The response of research and education to these themes will do much to determine how the university establishment is viewed generally. Below I will examine these themes in three modes:

1. what they are and why they will last;
2. how they relate to changes occurring and foreseen in forestry and environmental operations and in schools of forestry and environmental studies and the universities that harbor them;
3. what that may mean for Wisconsin and the University of Wisconsin System.

My working hypothesis is that latent public concern about the environment will be purposefully overt in the next generation, and that the current academic environmental presence will be wholly inadequate to answer the concern. At the same time much that is included in academic forestry and natural resources curricula and research (if agriculture was within my charge I would include it, but it isn't) has become moribund. Thus hard changes are in store for forestry and environmental academia, including growth, movement away from disciplinary and commodity focus toward solving politically and socially defined problems, and movement toward integrative research and teaching on a scale rarely contemplated now. The third is perhaps the most significant. The common cry of academics during the environmental ferment of the sixties and seventies, as I recall, was "But the environment is everything," with the clear implication that it was therefore nothing, at least in the context of universities as they then existed. Unless we learn this time to deal with "everything," we as educators may find our level of public and private credibility and support approximating nothing. On the other hand, we have a marvelous opportunity before us. We can take the strong elements from forestry and blend them with other, mostly available areas of knowledge and create a new profession to harbor journeyman environmental managers. Because "environment is everything," they will be found useful by society in a much greater range of activities than their predecessors. Indeed, because environmental management is a component of every human endeavor, they may become as plentiful as MBA's or even lawyers.

THE THEMES

Sustainability is the desire for an assured economic and cultural future. It doesn't mean doing the same thing over infinitely; it means having the materials and processes needed to change with the times and sustain the species. This especially includes what we call natural resources: our biological endowment of animals and plants, as well as air, earth, water and sunlight. The way we have taught about and managed these has not always been, and is recently seen not to be, supportive of sustainability. Sometimes this is a consequence of innocence, that is to say of an unwonted parsimony of thought and research.

As an example of this and of the sort of change in thinking and information that must fuel an adequate forestry response to sustainability, I take an example from the coastal forests of Southeast Alaska. There grow old stands of conifers all of about the same age. The question was how to harvest them to obtain a good stand of new trees, and how to keep the soil on the steep slopes, thus presumably to sustain the productive capacity of the land. Applying techniques from elsewhere, we decided that clear cutting, with little disturbance of the forest floor, would allow spruce to establish from seed, get plenty of light, and at the same time reduce soil loss and siltation. And that all happens. Current research is also showing that these practices may not sustain productivity. Wind throw is the natural method of forest renewal there,

Table 6.1

Nitrogen Self-Sufficiency for Silvicultural and Agricultural Crops

Forestry	Age (Years)	Nitrogen Sufficiency (Kg ha^{-1} yr^{-1})
Hardwood – natural	47	– 1.5
Hardwood – plantation	11	–16.2
Conifer – natural	53	2.5
Conifer – plantation	37	– 2.8
Agriculture		
Wheat	1	–79
Potatoes	1	–84
Corn	1	–168

and in the process, the soil is stirred and an impervious layer broken. Often on the mineral soil so exposed, as on sites logged with heavy disturbance in the past ("bad logging practices") shrub succession occurred that added nitrogen and readily decomposable organic matter to the soil. When disturbance is avoided, the impervious layer is not broken and the shrubs do not appear, and productivity may well decline. This sort of outcome is common where attempts are made to manage complex systems over long times with inadequate information and a too-narrow view of purpose.

Other times our failure to serve sustainability is the result of "rational" decision-making. We attempted to determine whether "intensive" management of forest stands usually resulted in enough of one nutrient element, nitrogen, remaining on the site following harvest to sustain another rotation at the current level of wood production (Bormann and Gordon, 1989). Often they apparently do not (Table 6.1). More alarmingly, we have apparently not been at pains to find out, although we commonly do economic calculations on wood production over a period of many rotations. The message for academia is clear: lighten up on the parsimony and reexamine rational decision making. The very phrase, "natural resources," may exemplify the problem.

Often, schools of natural resources are collections of advocates for the mental constructs we call resources, i.e. trees, wildlife, water, air, soil, even "recreation resources." This leads to the construction of academic discipline-like texts, courses and turf fights. Lost in the process are integration and the environment. Probably, the principle academic reason for having such a school on campus is to have at least one academic unit with no internal disciplinary boundaries, and an external problem focus. Only through integration will an academic base for sustainability be created, and then only if it becomes a university-level theme.

Global change is a fact long established, whether "global warming" is occurring or not (it almost certainly is or will). What is new is that the fact is coming widely to be believed and that a significant portion of the change is seen to be caused by us. We have not sufficiently incorporated change in natural resources into our teaching and research, nor have we communicated it to the public well. The notion that environments altered by human activity are necessarily worse than those unaltered or less altered is entrenched. Now that human effects are global, and perhaps dominant, in the environment it is time to examine in much more detail the theme of managed change. The fact that politicians like Thatcher, Gorbachev and Bush publicly declare themselves environmentalists recognizes that young people particularly have come to believe that degradation of the environment may affect them in their lifetime. No longer is it necessary to invoke "generations unborn" to get a hearing for protection of the environment. An entire generation of conservation rhetoric has been rendered obsolete. The human face of managed change is visible already in places like the Netherlands and parts of Connecticut. Its inhuman face is visible already in other parts of Connecticut and New Jersey as well parts of Amazonia and India.

Increasingly it is seen that environment and economic development are positively connected. Environmentally livable places attract productive people and modern industry. Perhaps the central challenge for forestry and natural resources people in academia and out, is to move their expertise to where the action is: economic development of an environmentally benign kind, in a society managed for change. This will mean reexamining old themes and attacking relatively new ones.

For example, if one were to imagine a sustainable economic activity with low environmental impact, one might imagine something very much like the manufacture of wood products based on conventional forest management. Things made of wood seem to retain their attractiveness in many markets; conventional forest management offers bountiful harvests in return for infrequent disturbance of ecosystems. Even then the disturbance overtly simulates natural ecological change. And yet, much environmental controversy surrounds just this set of activities. Perhaps one reason is that we academics have failed adequately to understand and explain the choices facing society. If this is so, it is

```
================================================================================
```

Table 6.2

Risk and Returns for Timber Investments, 1955-1985

	Louisiana	Pacific Northwest	Maine
Mean	0.056	0.071	0.053
Standard Deviation	0.151	0.171	0.107
Correlation with SP500	-0.024	0.178	-0.251
Capital Asset Pricing Model	0.224	0.384	-0.234
Inflation	0.044	-0.394	1.286

```
================================================================================
```

because our view has been too narrow and short-term. We have focused on controversies, localities, and quick fixes, and neglected to explain how choices about forests are related to long-term human needs and purposes. Often we have substituted platitudes for hard-edged scholarship. For example, we are just now getting around to explaining, in investment terms, why anyone might, or might not, want to invest in timberland. Still missing are integrated investment measures that cover resources other than timber (Binkley and Washburn, 1989) (Table 6.2). In particular, we have abandoned cities. The notion of landscape management and photosynthesis-based production is absent from most urban school curricula and most urban minds.

It is a short step from urban populations to "overpopulation" in the minds of most forestry and environmental people. Yet, human population has only uneasily been a subject in the natural resources community. It undoubtedly presents the greatest challenge to managers and academics alike in its complexity, immediacy, and reflexive implications. With more than 70% of the world population in the coastal zone, and with the threat of sea-level rise, distribution as well as magnitude is easily seen as an important subject. But population growth presents a cruel paradox: its limitation may be necessary to natural resource conservation and sustainability but limitation may only be achievable through increases in standard of living and attendant changes, such as improved female literacy. These in turn will only result from higher levels of economic activity that hasten resource degradation. What is the role of forestry and environmental academia in literacy, contraception and awareness of population changes and effects? Can forestry knowledge be used to resolve the economic development/population limitation paradox? The picture is clearer but not terribly better served currently in fitting people into the environment. It is now clear that the idea of preserving biological diversity and "nature" is not sufficiently accomplished by simply establishing parks and preserves. Ecosystem or landscape management in which the human neighbors and inhabitants of parks are integrated has much promise, but is an infant science for which we have not created a sufficient knowledge base (Gordon, 1989).

Nor is the picture better from a pure resource or commodity view. Forest acreage per human is declining worldwide, as opposed to an increasing world demand for wood products and the environmental benefits of forests. Ocean fisheries decline, and agriculture needs a major overhaul.

TRENDS IN PRACTICE AND EDUCATION

What is happening in the forestry and environmental education community currently in response to these issues and conditions? According to Giese (1988) both human and monetary resources for forestry research are declining. Both federal and industry allocations for forestry research are down in real terms over the last decade. In addition, the number of students entering forestry programs is declining, thus decreasing the current need for forestry faculty, and decreasing the pool from which future researchers will be drawn. Giese lists six consequences of the decline in the forestry research system:

- We are suffering from the loss of synergy that results from teams of diverse scientists working to solve large-scale and long-term problems or smaller problems that are specific to individual species. The personal (and often unpublished) knowledge of interactions is being fragmented; i.e., some of the "art" is being lost through forced career changes and early retirements. Knowledge gained through negative results of experiments, which is usually not published, is being lost through cutbacks and retirements. The natural consequence is that in the future we will surely find ourselves redoing experiments that have been done before, but which could have been avoided had there been opportunity for conventional communication among scientists.
- We are forgoing long-range, bench mark projects of the Hubbard Brook type (an ecosystem-level study, now about 30 years' duration, to elaborate on the effects of forest management on nutrient cycling) by providing funding for trendy research. Not all research should be in biotechnology. There are many other important biological and social research problems that need different approaches and a sustained effort.

- New, difficult problems (including acid deposition, international competitiveness and tropical deforestation) are emerging, but we have hardly begun to understand them, let alone prescribe solutions. Again, a sustained effort is needed.
- We don't understand how to manage our forest resources for multiple use. Use increases, but our knowledge of the cumulative long-term effects has not kept pace.
- We have come to realize that we are part of a global economy and ecology. The implications of that fact are being felt even at very local levels. Yet local communities do not have the resources or expertise to anticipate these global trends or to develop strategies for dealing with them.
- We cannot look to the sister sciences of botany and biology for assistance, because scientists with backgrounds other than forestry will hardly fill the gap; perennial plants are seldom the species of choice for biological science.

The stark contrast with societal themes is manifest. Duncan et al., (1989) examined forestry education and its likely impact on the future of the profession, and described several problems with the public image of forestry related to forestry education as now delivered:

1. Forestry is viewed not as a profession but as a technician-level vocation requiring the ability to fight fires, fell trees, oversee logging crews and drive tractors—a low-tech image with limited educational requirements.
2. Foresters prefer to work with data and things rather than with people.
3. Although there is a range of forest values, foresters are primarily interest in timber.
4. Foresters are essentially applied biologists.
5. Forestry is just one facet of agriculture.

They conclude that forestry education as a whole is deficient in self-examination and strategic planning, and that this has lead to a decline in student quantity and quality, a poor record of including women and minorities, and poor integration with other natural resource and environmental academic units. Again, the contrast between the negative trends in forestry education and societal needs are striking.

Are changes underway that will begin to match forestry education to societal needs? The appearance of frank appraisals of the kind that Giese and Duncan et al., present is a good sign. The National Academy has recently commissioned a study of the future of forestry research, and some institutional change is apparent. The injection of social science into forestry curricula is gaining favor. Graduate professional degrees are gaining as first professional degrees, and these are drawing a broader spectrum of students into forestry and environmental management (Gordon, 1984). Thus the broadening of forestry curricula and the creation of a more diverse student population, given a higher level of professional instruction, may improve the match between academic forestry and societal needs.

Forestry and agriculture are more often found together, at least administratively, than apart on American campuses; although it is interesting that several old and prestigious forestry schools are in universities where agriculture is weak or absent, for example, Duke, Michigan and the University of Washington. It may be important for forestry and environmental management to be clearly differentiated from agriculture and then reintegrated with agriculture and other disciplines into a science of landscape management. Although similar people and concerns have often marked the history of agriculture and forestry, they have important fundamental differences in their classical academic form (Gordon and Bentley, in press). Thus their integration is not an easy or automatic task, although it is often viewed as an accomplished fact.

Nowhere has this become more painfully evident than in the attempt to launch "agroforestry" as a research discipline in aid of rural development, usually overseas. Forestry schools, although several have been recently quite active, have usually lagged agriculture in the internationalization of their teaching and research. Internationalization cum globalization is now seen as a major and urgent need on U.S. campuses if we are to remain competitive in a shrinking world. Forestry has a long way to go before curricula are global, and the regional or state focus of many forestry schools is seen to conflict with this goal.

Perhaps the most urgent and underdone task facing forestry and its academic allies is the integration of urban and rural landscape science, once they are created. Urban spaces and the "built environment" are often viewed as devoid of natural resource management need and as lying outside an ecosystem context. However, the setting and maintenance of buildings, statues and monuments, and the realization of artistic, social and engineering goals in cities are dependent on environmental variables and can be aided by ecosystem and forestry concepts. Perhaps the most urgent task facing forestry is the diffusion of environmental capability and awareness throughout society, particularly to urban populations that are politically potent and without direct connections to the country. Opening of forestry and natural resource schools to the rest of the university is the first step. Too often, only one or two "survey" courses are available to non-majors, and thus the larger student population of the university, most destined to be urban-dwellers, are insulated from forestry concepts. Perhaps as forestry schools, or what they become, move toward graduate professional degrees, faculty formerly occupied with teaching forestry undergraduates can offer courses more broadly. This in turn may produce more top liberal arts graduates as feed stock for graduate professional programs. Also it should aid the installation of environmental courses and the enhancement of science in the liberal education paradigm. Ultimately, by this process, forestry may no longer be seen as dirty or arcane, nor environmental studies as flaky, by the average graduate.

This in turn may help gain for forestry and the environment (and agriculture if they choose to play) a firm place in primary and secondary school curricula. Give students something worth learning and they will learn it. Academic skills can be practiced in any subject since they are reading, writing, speaking, counting and thinking. Thus, they can be practiced and sharpened on the environment and natural resources, which are attractive and useful subjects. In this way, forestry and environmental higher education may contain part of the solution to primary and secondary school reform. Don't teach math, teach environment in mathematical ways.

To capitalize on these trends and opportunities, forestry, the environment and natural resources must emerge as university-level themes. Academic administrative units at the core of the university level effort may well emerge as collections of specialists in fairly large groups without departmental or disciplinary walls, and with strong ties to other parts of the university. Forestry and environmental studies may thus provide the university with a useful comprehensive model: disciplines focused on problems and opportunities in flexible arrays, rather than departments focused on disciplines in rigid, timeless ranks.

IMPLICATIONS FOR WISCONSIN AND THE UNIVERSITY OF WISCONSIN

Wisconsin perceives itself as an environment-dependent state, has a leading research university, and should play a leading role in academic change to meet these challenges. The invention and propagation of landscape science should be the new land-grant mandate and Wisconsin and the University of Wisconsin are good places to forward it. The forest products industry, tourism, rural development, and urban-rural integration will all be important to Wisconsin's economic and spiritual future. Indicated changes have begun to occur as this conference attests. Thus, I list below several actions to consider should this course be chosen:

1. attract faculty integrators who can cross disciplinary boundaries in research, and who are interested in teaching a broad audience;
2. greatly increase outreach efforts to traditional clients and redouble efforts to get into the primary and secondary schools;
3. institute landscape-level research that integrates the testing of hypotheses about social and natural science questions. Make the creation of ecosystem management philosophy and practical methods the highest research priority;
4. concentrate teaching at the undergraduate level on the university at large and on the public. Make the masters degree the first professional degree; rely on other state institutions for the production of bachelor-degree foresters;
5. focus the graduate professional curriculum on the science base, communication, and the management of people and money. Provide hands-on skills through field courses and internships.

By moving in these directions, Wisconsin can lead in addressing the urgent and persistent American and global themes of sustainability, global change and population. Only by attracting and educating people capable of integration on a grand scale, who are at the same time practical and client-driven, can universities do their part in sustaining the human enterprise. Although much remains to be detailed, I believe the path is relatively clear. Do we have the institutional will to follow it?

BIBLIOGRAPHY

Bormann, B. T. and J. C. Gordon. 1989. "Can Intensively Managed Ecosystems Be Self-sufficient in Nitrogen?" *Forestry Geology & Management* 29.

Binkley, C. S. and C. L. Washburn. 1989. "Financial Risk Differs By Timber Regions." *First Watchovia Timberland Investment News* Second Quarter.

Duncan, D. P., R. A. Skok and D. P. Richards. 1989. "Forest Education and the Profession's Future." *Journal of Forestry* September.

Giese, R. L. 1988. "Do Declines in Research Funding, Business Profits and Student Enrollment Spell Trouble for the Profession?" *Journal of Forestry* 86(6):15–22.

Gordon, J. C. 1989. Recommendations of the Commission on Research and Resource Management Policy in the National Park System. Trans. 54th North American Wildlife & Natural Resources Conference 1989:12–14.

7

Rural Land and Natural Resources Management

MARK B. LAPPING

It is increasingly apparent that we can no longer speak of land and resource management issues as if they were or could be separated from international economic, social, political, and environmental pressures and constraints. Indeed, the globalization of the economy and the biosphere are among the most powerful realities with which any exercise in strategic planning must contend. Global and even national assessments, however, tend to overwhelm us, both by their sheer immensity and scale as well as the ferocity and pace of change which defines our very existence (*Scientific American*, 1989).

We now live in a world in which a catastrophe like Chernobyl not only reduced its immediate impact area to a poisoned wasteland but also had the capacity to radically transform man/environment relations and the unique social fabric of a far-away indigenous people, the Saami (Lapps). What Fenno-Scandinavian colonization could not do in a millennium, a nuclear accident did in a matter of days (Cline, 1988). We now live in a world in which acid precipitation and the deposition of PCBs, mercury, and toxophenes originating in one region alters the waters, soils, wildlife and forests in others. We now live in a world in which the ozone layer is affected adversely by the billions of incremental emissions from spray cans in daily use everywhere, not to mention the impact of the automobile.

We now live in a world in which the destruction of tropical rain forests is changing the climatic expectations of whole regions with the result that agricultural systems are being put in jeopardy. We now live in a world in which entire species are liquidated in a routine fashion, syringes cover the beaches, Exxon Valdez–type spills and Bopahls occur with alarming frequency, precious soils continue to erode away, and garbage barges roam the seas in search of a destination. All of this is witness to the fact that, as Lewis Thomas has so wisely observed, "we are now the dominant feature of our own environment" (Brion, 1988). We have yet to come to a full understanding of what this means for us.

Local assessments of our collective situation, on the other hand, have about them a sense of incompleteness in that exogenous factors play ever larger roles in our near and home environments. Further to the point, we are coming to appreciate the shortcomings of local decision making processes which cannot, almost by definition, confront problems whose origins lie beyond our boundaries, do not bestow property rights and stewardship obligations, and for which prices and markets are nonexistent. Still, we have little recourse but to reflect on the wisdom of Rene Dubos' injunction that we must think about such things globally, and act locally.

In all of this Wisconsin is a part. The future of this state and its people is inextricably bound-up with the global reality in its many manifestations. To act as if this were not the case is to contribute further to the self-delusion and arrogance to which Thomas has alluded. To act as if this were not the case is not only absurd but surely counterproductive. Thus, my comments will seek to tie together some larger themes to those matters closer at hand. This is as it must be because Wisconsin is one of those places whose future relies so heavily upon plans, decisions, and policies made about rural lands and natural resources.

RURAL LAND AND NATURAL RESOURCES

Wisconsin's economy, like that of the nation as a whole, is in transition from manufacturing and natural resource development (mining, forestry and agriculture, for example) to a greater reliance upon services. A steady decline in the proportion of the workforce employed in traditional sectors of the Wisconsin economy remains unabated while growth in the services category continues (Wisconsin Department of Development, 1987). However, what is not always obvious in gross and crude sectoral analysis is that all sectors of the Wisconsin economy depend, to some extent, upon the health, integrity and quality of the Wisconsin rural land and natural resource base. This will continue well into the future, even as sectors change in their relative importance as a percentage of total employment and the state's gross product.

For example, Wisconsin's manufacturing sector includes important small engine, mining and agricultural equipment production and sales, all tied to rural-based demand. Likewise, the state's huge forest industry— Wisconsin ranks first among all states in the production of writing paper, specialty papers and sanitary tissue products—is highly dependent on Wisconsin's forests, both public and privately controlled. Indeed, as one study points out, "Wisconsin's nearly 1,450 forest products companies are a mainstay of the rural economy, employing 83,000 residents and producing $9.8 billion worth of products annually" (Wisconsin Department of Development, 1988).

In agriculture, cash receipts in 1984 exceeded $5 billion. Fully 18 million acres, or slightly more than half of the state's entire land base, are currently devoted to agriculture on the state's 80,000 farms (Department of Agricultural Economics, UW–Madison, 1988). Approximately 100,000 Wisconsinites derive their livelihood directly from farming with countless others employed in the large agribusiness sector which is so dependent upon the quality and quantity of foodstuff generated by Wisconsin farmers (Wisconsin Department of Development, 1988). Not only is Wisconsin preeminent in dairy products, but it is also a major national producer of field crops, corn for silage, cranberries and even maple syrup. Food processing, a vital "value-added" part of the agribusiness sector, is especially significant in the western counties of the state.

Even in the services sector, natural resources—rural waters and rural lands—are central to the burgeoning hospitality-recreation-tourism industry. This portion of the service sector is considered by many to have extraordinary growth potential (Wisconsin Strategic Development Commission, 1985). In 1985, the recreation-tourism subsector accounted for over 11% of the state's total employment (173,792 jobs) and generated approximately $14 billion to the gross state product (Heberlein, 1987). Management and planning decisions about Wisconsin natural resources and Wisconsin land will have a profound impact upon the future of this sector. While tourist-based industries are important to metropolitan Wisconsin, they are vital to the relatively poorer and more rural northern tier of the state. Recent in-migration into these regions is very largely tied to the resource dependent recreation and retirement subsectors. In Vilas County, for example, it is estimated that fully 60% of all personal income is derived from persons 65 years or older. The majority of these residents have remained or migrated to the area because of its outstanding resources and amenities (Wisconsin Department of Development, 1988).

Farmland Preservation and Ownership

A large number of issues and vulnerabilities exist relative to Wisconsin's future land base. Perhaps paramount among these is the widespread concern over the diminution of farmland, a major theme in Wisconsin for the last twenty years or so. Most states, and the federal government, have instituted a number of policies and programs to both assess the nature and magnitude of agricultural land loss and to arrest the loss, where that is possible. Some students of the problem have argued that the loss of farmland, together with an array of land degradation problems, has only intensified as a national and international problem. Others have argued to the contrary. In a recently published paper, for example, Pierre Crosson and Norman Rosenberg (1989) have concluded the following:

> In the 1970's a study by the U.S. Department of Agriculture and the Council on Environmental Quality concluded that by the 1990's, the conversion of agricultural land could present the nation with a resource-scarcity problem as serious as the then prevailing energy crisis. Subsequent analysis has greatly diminished such concern, and the issue is no longer much discussed in the U.S.

Such a cavalier dismissal of the problem of farmland loss and farmland quality is potentially dangerous. Further, Crosson and Rosenberg's assessment of the matter fails to treat seriously the nature of these issues on the state and local levels. Of this problem, noted agricultural economist Philip Raup (1980) has said:

> the highly aggregated nature of the data masks the impact of the steady loss of cropland on specific regions and localities. It is the irreversibility of the conversion (from agriculture to more intensive, urban uses) and not its magnitude alone, that provides the strongest root of concern over the competition for land.

The fact of the matter is that farmland is simply not farmland. Some lands frankly are more valuable than others because they have unique production capabilities as well as versatility. Not all soil, micro-climates, or hydrological systems are alike. An acre of land in eastern Kansas can do a few more things than an acre of land in central Kansas, for example, but the loss of neither of these acres is the same as the loss of an acre of land on Long Island to New York City. The land on Long Island has unique characteristics that make it singly important, not only in terms of the factors of community life, but for the stability of agriculture. Clearly, in many agriculturally defined areas the loss of farmland is a very critical issue, but for the larger community, the loss of unique and regionally appropriate agriculture is the greater concern.

Farmland trends in Wisconsin, with some variation, have tended to parallel national trends. In the mid 1930s, Wisconsin had approximately 200,000 farms. Today the number is approximately 80,000. Many of the farms which have gone out of production were marginal in their productivity and capability. Some were located in the northern parts of the state, areas which, some would argue, should not have been cleared for agriculture in the first place. Still, as Barrows and Barlowe (1979) have pointed out, "significant areas in each state (including Wisconsin) shifted to urban and urban-associated uses."

Wisconsin has sought to deal with this problem through an imaginative set of strategies. Starting with property tax relief for farmland owners in 1974, the program has matured to include tax relief tied to local agricultural zoning and comprehensive local planning. The results have been impressive. As one study (Jacobs and others, 1989) has noted:

> [The program] by 1988 appeared to be a huge success. Over 8,000,000 acres are enrolled under the program, representing 50% of eligible farm lands. All but two of the state's counties have a certified farmland protection plan, and 361 of the state's 1,268 towns have certified zoning ordinances.

Another assessment calls the Farmland Preservation Program "one of the most notable programs which has benefited the agricultural community. . . ." Moreover, the program "provides approximately $30 to $35 million in payments annually to qualifying farmers and offsets about 10% of farm property taxes statewide and 38% of the property taxes of farmers who claim the credit" (Wisconsin Department of Development, 1988).

Just how long Wisconsinites will provide such relief and will be willing to absorb such a re-allocation of the property tax burden depends, to some extent, on the returns they believe they will receive from such approaches. It is not altogether clear that local political will and support for such programs will last into the future. Indeed, given the shrinking number of agricultural-dependent counties in Wisconsin, as elsewhere, some degree of concern must be expressed. This is further exacerbated by a general ignorance of those priced and non-priced goods and services produced by agriculture, i.e. habitat, air quality, etc.

While the amount of land devoted to agriculture in Wisconsin has declined, the size of the average farm has increased very substantially. In 1950, for example, the number of farms working fewer than 100 acres was approximately 85% of the total. This number declined to 53% in 1982, while those farms working between 100 and 199 acres of cropland doubled during the same period. The average size of the family farm in Wisconsin has increased from 138 acres to 210 acres during the period of 1950 to 1982 (Novak, 1985). This and other factors indicate a growing level of farm concentration and intensification throughout the state.

Together with commodity specialization, such concentration and intensification—accompanied as it usually is by heavy applications of nutrients and pesticides—tends to increase the potential for erosion and pollution. Further, a trend toward the utilization of more rental lands for agriculture raises further the concern over on- and off-farm impacts of intensification. The evidence on rental lands is convincing: such pieces of farmland tend to be farmed in a more intensive manner than are owner-operated acres. A reliance upon such lands—which now constitute one out of every four acres in production in Wisconsin—suggests that erosion and pollution will also increase if precautions and best-available technologies are not employed.

Changing land tenure patterns, then, must also be a concern for those with an interest in the future of the farmland resource. In New England, for example, only one state's farmers own the majority of the farmland resources; that state is Vermont where 54% of the farmland is owned by farmers (Lapping and Clemenson, 1983). If you talk about farmland in other New England states, you need to talk about a few insurance companies, banks, and investment firms. In other words, we are looking at an industry where the old notion about independence and self-sustainability has really flown out the window. If farmers no longer control such assets, we will continue to experience dissonance between those policies aimed at conservation and pollution control and those which seek to maximize agricultural productivity at any cost.

A tenure issue of who controls the resources is going to be profoundly important. Who will be the next generation of Wisconsin farmers? Where will they come from? Will there be intergenerational transfer of assets? How are they to be trained, nurtured, and educated? Where are they going to get the intelligence of local agriculture which necessarily is not diffused by agricultural institutions?

Because of variations in soil, hydrology, and climate, local intelligence of agriculture is important. One of the things I discovered in editing a regional history of three centuries of farming in New England was the story of wheat in Vermont. In the 1860s, the Champagne basin was one of America's primary wheat production areas. The wheat blight and the settling of the Northwest Territories began to shift production to the Midwest. At the start of World War I, the U.S. was desperate for wheat, and for all sorts of grain production. The federal government offered $1.00 per bushel incentive over the going price. No one in the Champagne basin remembered how to grow wheat. In fifty years, the intelligence of agriculture dies. Maintaining our understanding of regional and local variations is learned walking up and down the land with a master farmer; it is part folklore and part academic learning.

The farmland preservation issue has still other important local dimensions. As lands are converted into urban uses, a new set of problems emerge for those remaining in agriculture. Some have noted that an "impermanence syndrome" takes over with farmers, who believe that land markets have changed in such a way as to lead to the eventual conversion of their land, ceasing to make necessary and appropriate investments in their operations (Lapping, 1979). When updating and investments no longer take place, productivity and stewardship lag and deterioration sets in. In a sense, this leads to a situation in which conversion becomes a "self-fulfilling prophecy." Fear of conversion leads to an unwillingness to invest further in agriculture which, in its turn, creates a less viable local agricultural economy that stimulates the movement of agricultural land into other uses, as profits decline relative to anticipated alternative gains.

As the number of farms dwindle, so does the "critical mass" for local agriculture (Dhillon and Derr, 1974; Conklin, 1969). As fewer and fewer farms remain in production, new strains are created for the local agribusiness sector. Firms, markets, and institutions are placed in jeopardy and this has the tendency to further erode the viability of farming. What is most crucial about "critical mass" is the need to protect entire agricultural subregions, not just random farms. A program of land retention must be based on the understanding that minimum numbers of units are necessary to support a thriving agricultural industry, which in turn helps to keep farming economically viable.

When people talk about Europe, they very rarely reflect on the fact that the Swiss government pays farmers to keep housed up on the alpine meadows. And, it is not because they need Swiss chocolate. It is because they know that tourists like to see those beautiful, open meadows. It is a fundamentally honest way to deal with farmers. Farmers do many things besides grow food. They protect the airshed, watershed, wildlife habitat, scenic beauty, and quality of life. The Swiss system is capable of paying farmers for being guardians over the historic and touristic landscape.

In the U.S., we do not pay farmers for growing food, let alone for other things. My father, who is hands down the smartest man I have ever met, says "Well, you know Mark, nothing is going to keep this farmland in production more than a couple more dollars in my pocket." If we could move away from the notion that farmers, like social workers, school teachers, and ministers, ought to pay us to do what they do, farmers might be able to retain more land in agricultural production. My father daily milks 72 to 74 head of Brown Swiss, a beautiful, registered 25-year investment. He makes more money selling maple syrup than he does from dairying, even though he tapped into the Ontario-Quebec market where the high butterfat milk goes into the pastry industry.

Along with the lack of economic viability of agriculture is the whole issue of critical mass. Every time we lose one farm, two farms, or three farms, we began to erode the critical mass. We must move away from the belief that farmers are warmed over neo-Jeffersonians, or like the self-reliant independent Waltons where everyone takes care of everything on Waltons Mountain. Farmers, like foresters, depend upon a unique, highly-integrated, and independent support system. There must be seed salesmen, there must be vets, there must be equipment sales. Everytime we lose farms, we reduce the viability of the support of the agribusiness system; every time we cut down on the viability of the agribusiness system, we make farmers and farm production that much more vulnerable.

Nuisance Claims and Right-to-Farm

There are emerging a number of pressures exerted on farmers to discontinue operations due to the adverse off-site effects they are perceived to have on neighboring land uses. These "right-to-farm" problems occur when a lot of people come to live in rural America who do not have an association with agriculture. Many of these folks seek farmland without farmers. When you live on Skunk Hollow Road after a time you notice that your BMW has axle problems because the road isn't paved. When my father moves his cattle from one pasture to another across the road, manure gets on the road because the cattle don't have diapers yet. Farmers operate machinery on Sunday mornings. Farm machinery gets in the way numerous times. One would prefer that there were no smells and that there were no dangers. Thus, increasingly farmers are facing complaints from the rural non-farm population that often do not understand that farming is a business—a very dangerous business for farmers and sometimes for others.

Collectively, these have come to be known as "right-to-farm" or nuisance problems; they are being addressed by "right-to-farm" laws in several states, including Wisconsin. Although they vary considerably, right-to-farm laws attempt to do two things. First, they all seek to supersede the common law of nuisance, the fundamental area of law used to challenge farming in the peri-urban areas. Second, they attempt to favor agricultural uses of the land above all others, especially those that are inherently competitive.

Right-to-farm laws seek to establish a "first in time, first in right" logic wherein prior farming uses of land have primacy over all others. These laws are founded upon the idea of altering the common law doctrine of nuisance to protect existing farming operations from conventional nuisance claims. The evolution of the nuisance doctrine over the years has been a tortuous one. In fact, it has been stated that "[t]here is perhaps no more impenetrable jungle in the entire law than that which surrounds the word 'nuisance' " (Prosser, 1971).

Right-to-farm laws, attempting as they do to reorder the relative property rights of neighboring land users, are concerned primarily with the private nuisance. A private nuisance is a civil wrong whose remedy lies in the hands of the individual whose rights have been disturbed. In order to have a nuisance for which the law will provide a remedy, there must be a substantial and unreasonable interference with the property interest being asserted.

This interference can be either negligent or intentional. For the purposes of farm operations, an action constituting a nuisance is deemed to be intentional, even if it is unintended, when it is a foreseeable consequence of the farmer's otherwise protected farming activities. Thus, the drifting of sprayed farm pesticides onto a neighbor's land is considered an intentional nuisance, even though this particular result is unintended (Mandelkar and Cunningham, 1985). A nuisance emanating from farming operations may also arise from negligent conduct, where the farm operator has failed to take appropriate precautions against risks apparent to a "reasonable" man or woman. The right-to-farm laws generally except from protection agricultural activities that are conducted in a negligent manner.

Most nuisance-driven land use disputes focus on the reasonableness of the defendant's conduct. Since all property owners are entitled to the reasonable use and enjoyment of their lands, some balance must be struck between the discordant, and often incompatible, uses to which the lands are being put. In each case, the court must make a comparative evaluation of the conflicting interests according to objective legal standards, and the gravity of the harm to the plaintiff must be weighed against the utility of the defendant's conduct (Prosser, 1971). Thus, something approaching a "balancing act" often takes place upon judicial review.

Because many otherwise reasonable actions can be considered nuisances under particular conditions, courts are often hesitant to find fault under the nuisance doctrine. "Liability is imposed only in those cases where the harm or risk to one is greater than he ought to be required to bear under the circumstances, at least without compensation" (Lapping and others, 1983).

Right-to-farm laws tip the balance further in favor of the farmer, by statutorily declaring that standard farming practices are reasonable land uses, despite their potentially adverse impacts upon neighboring lands. The laws also alter the balancing process by establishing, legislatively, that the utility of farming outweighs, at least to an enhanced degree, some measure of incidental harm to neighboring landowners. Right-to-farm laws also modify relative property rights in instances where property owners are said to have "come to the nuisance." This oft-litigated aspect of nuisance law involves situations where an aggrieved plaintiff/landowner has purchased and occupied land despite the presence of obviously incompatible land uses.

At the heart of right-to-farm laws is the desire to protect innocent farmers from land use actions or conditions that evolve around them, over which they have little or no control. The possibility exists, however, that the land use conflicts precipitating a nuisance lawsuit were caused by the farmer himself, as he shaves land off the farm and sells it for residential development. Only Washington's law addresses this issue directly. That state's law offers no protection under the right-to-farm principle to a farmer who has promoted land use change through severance or subdivision. This will be a matter of future concern elsewhere as once predominantly agricultural areas become mixed in their land use and spatial patterns and farming interests are perceived to be creating or contributing to the situation.

Right-to-farm laws do not address the real threat to farmland from the land's increased value and the potential economic gains available by converting farmland to more intensive uses. This is the issue facing many farmers who operate in the city's shadow. Further, even the limited effect of right-to-farm laws is threatened by the possibility of suits in trespass. These actions could attack the very conditions that are legalized under the protection from nuisance actions provided by the right-to-farm statutes. Ultimately, the courts will determine the effectiveness of the right-to-farm laws. However, these statutes are—with the possible exception of differential/preferential tax schemes—the most popular form of agricultural land or land-related statute on the state level. This may reflect the inherent fears all farmers have in dealing with a general population that had grown ever more alienated from the larger realities of agriculture.

Alternative demands on agricultural land are increasingly being generated by the very same governmental units that are also pursuing the protection of farmlands. The siting of "high risk" land uses has given rise to its own vocabulary. We now speak of "LULUs" ("locally unwanted land uses") and "NIMBY" effects ("not in my backyard"). Waste treatment facilities, prisons, disposal and sanitary landfills, and other such "nuisances per se," must be cited. These are not luxuries but critically important facilities. As citizen groups, municipalities and other interests successfully resist the location of such LULUs in their locales, governments are forced with greater frequency to situate such facilities in rural areas. This will often mean that farmlands will be purchased for such uses or will be condemned. The ease with which government may do this to agricultural land depends, in part, upon the nature and record of state policy vis-a-vis agricultural land preservation, as Margaret Grossman (1985) has demonstrated.

Competing Land Uses

While a "hard and fast" rule does not yet exist, there appears to be a strong correlation between the degree to which a state seeks to protect farmland and the ease that governments will have in acquiring farmland for other public uses. In those jurisdictions where the goal of farmland preservation has received little attention, the demonstration of another compelling "public use" is a relatively easy matter. Conversely, in those states where the protection of agricultural land is a major policy concern, it appears that a more rigorous review of the "public use" criteria will take place. In Kansas, where farmland preservation is not perceived as an important issue, the public benefit of a condemnation for another use of the land is a matter of little dispute, except for the affected landowner. But in a state like Wisconsin where major efforts have been made to preserve farmland, the demonstration of a compelling "public use" will be a more difficult matter to prove (Lapping, 1990). Still, few can doubt that such conflicts for rural lands will not increase.

Perhaps the ultimate paradox relative to farmland lies in changes in the structure of agriculture. In Wisconsin, as elsewhere, employment in non-agriculturally related sectors has come to typify many rural economies. The dependence upon manufacturing is especially noteworthy. One Wisconsin study suggests that well over 85% of the rural "working" population is engaged in non-farm pursuits (Wisconsin Department of Development, 1988).

Both national and local studies attest further to the fact that farm families are increasingly relying upon off-farm incomes as farm-generated revenues continue to decline. Glen Pulver and his colleagues have shown that this is a growing trend among Wisconsin farmers (Department of Agricultural Economics, UW–Madison, 1988). In a very real sense, farms and farmers are becoming increasingly dependent on non-farm income inputs. This may mean, in essence, that to secure a successful and viable farm economy it will become necessary to "grow," attract and expand the non-farm employment options in rural areas. Without adequate and sophisticated planning, such strategies will necessarily increase competition for land between farmers and non-farm users. This issue will become more clear and will exacerbate with time.

Land Degradation

Compounding any discussion on agricultural land use is the issue of land degradation. Indeed, the two matters are inextricably joined. As agronomist Frederick Swader (1980) has written:

> American agriculture has unwisely used the soil and allowed soil erosion and compaction to reduce the stock of soil available for future farming activities. When combined with the fact that an estimated million acres of cropland are converted to other uses annually, there is growing concern about a long-term inability to meet domestic and world food demar.ds at affordable prices.

Again there are those who minimize the impact of soil erosion by pointing to the recent history of increasing yields per acre brought about by the incorporation into agriculture of advanced technological innovations. Yet some have warned that the sustainability of such innovation and diffusion is a matter open to debate (Crosson and Rosenberg, 1989).

Salinization, soil erosion, water pollution and other forms of resource degradation have been observed throughout Wisconsin's agricultural system. Erosion alone appears to be a vast problem. As Novak (1985) has observed:

> Erosion from cultivated Wisconsin cropland in 1982 totaled about 64 million tons of soil. The average erosion rate from cultivated cropland was 6.8 tons per acre per year. This compares with most Wisconsin soils' tolerance value of four tons per acre per year. . . Wisconsin is exceeding its T-value on cultivated cropland by about 70%.

Even if one were to discount for regional variation, this is astounding. No one seriously concerned with Wisconsin's future—let alone its agricultural viability, soil and water quality—can doubt the challenge this presents for all sectors of society. Once again figures in the aggregate may suggest to some that nationally this may not be a profound dilemma. But as Wisconsin's fertility blows and floats away can anyone seriously doubt the local and statewide magnitude of the issue? One would think not.

Forestlands: The Multiple Use Debate

The other critical rural land resource is the state's forestland base. As in many other states, the private non-industrial woodland owner (PNIF) controls many of the most significant forestland resources in Wisconsin. Approximately a third of the state's private forestlands are part of a working farm with farmers reporting wildlife habitat, on-farm utilization and scenic enjoyment as significant management objectives. In the southern part of the state pasturage is another frequent use of such farmer-owned forestlands (Roberts and others, n.d.). Much of the remaining PNIF is controlled by others with varying interests in the land; over a fifth of the total is controlled by retirees. Altogether a recent Wisconsin survey has indicated that there are nearly 220,000 woodland units in private non-industrial ownerships and that over 9 million acres of forestland fall into this category (Roberts and others, 1986). Slightly over 15 million acres statewide are in forestland, with Wisconsin's significant forest products industries owning less that 10% of the total (Stier and Jordahl, 1987).

Since PNIF owners control the overwhelming majority of the state's forestlands, their objectives, management approaches and investment strategies are critical to the future of the state's forestry industry and stock of environmental resources. Perhaps not surprisingly, the production of timber does not appear to rank very high as a goal among these owners. Rather, "scenic enjoyment" and "wildlife habitat" are the two most important reasons for ownership, according to recent surveys (Roberts and others, 1986). Wisconsin has had a long tradition of public policy directed toward support for the PNIF, starting with the 1927 Forest Crop Law and culminating recently in the 1985 passage of the Managed Forest Law (Tlusty and Jordahl, 1988). Careful evaluation of this program will be required to determine its effectiveness.

To the extent that amenity values are paramount objectives for PNIF, the state's growing recreation and tourism subsectors are likely to enjoy a relatively secure land base. This is especially true given the very large number of forestland owners who allow access on and across their lands to recreationists. But other issues exist, not the least of which is that of the continuing parcelization of substantial blocks of forest into ever smaller units. John Roberts (1987:22) has pointed to a number of potential issues generated by parcelization:

> Evidence suggests that owners of smaller woodland parcels are less likely to have harvested wood products [for sale], seek or use professional advice regarding management of their woodlands, intend to harvest wood products [for sale] in the future or permit public use of their woodlands for recreation. . . . The long run implications of all this are obvious.

The ramifications of such a forestland ownership pattern for the state's forest industry must be examined and clearly understood. Experience in other states suggests that serious issues for management and efficient plant operation have emerged in those places where parcelization has greatly increased. Further, "as ownership shifts from farmers to nonfarmers and into smaller size ownership, purely economic incentives, at present levels, may be attractive to even fewer owners" (Roberts, 1987). The future policy direction may require a move away from its historic reliance on such approaches.

As with agriculture and recreation-tourism, Wisconsin's forestry sub-sector must have additional product and marketing opportunities if it is to retain its capacity and remain highly viable. Issues related to acid precipitation, soil erosion and degradation, and forest species quality and diversity are as important as any facing the state. Forests are incredibly productive and resilient. They provide an array of priced and non-priced goods and services. As they constitute one of the state's most important land uses, they will increasingly merit attention and investment. To do otherwise would be amazingly short-sighted.

CONCLUSION

Time, space and other constraints have prevented a more thorough discussion of some of the truly significant "Great Lakes" issues, such as the massive loading of anthropogenic toxic chemicals in the Lake Superior ecosystem, changing energy supplies and costs, climatic change, and global population realities. Those issues must remain for another time and perhaps other commentators.

One of the central notions expressed in any strategic planning process is that it is critical to concentrate on those things that are susceptible to change and direction, as opposed to those things that are beyond one's power and ability to control. The recent report of Wisconsin's Strategic Planning Council (1989) has said that:

> Wisconsin's economic destiny continues to reside in Washington, Tokyo, London, Bonn, Rome and Paris. External factors and events will have the most significant impact on the future of Wisconsin's economic recovery.

True as much of this may be, how Wisconsin will address its significant rural land and natural resource issues is a matter very much in the hands of Wisconsinites and their institutions. Clearly what is requisite is a conceptualization and rationale for sustainable development with environmental protection which can be operative and effective. Part of the essence of such an ethic exists in the long and rich history of land and conservation policy that is this state's great legacy (Jacobs and others, 1989). Part of this can be found in some of the state's pathbreaking judicial record, as in the seminal case of Just v. Marinette Co. (56 Wisconsin 2d 7. 210 N.W. 2d, 761, 1972), that changed forever the definition of the state's responsibility to safeguard environmental quality and natural resource integrity (Large, 1973; Lapping, 1975; Brion, 1988). And part of this can be found in a modest reformulation of the "Wisconsin Idea," a set of values that made this state's university system among the most thoroughly important, relevant, and socially responsive in the country.

EPILOGUE

Let me close with five nasty comments about universities and about all of us. First, we tend to talk a great deal about a work ethic, but we do not respect work. Anyone who does not wear a white collar, or recently a pastel collar, does not work. We do not value that work. Our concern is not mirrored in other countries. In Sweden, farmers are entitled to a wage, or return on investment, equal to someone who works at Volvo, but that is not true in America. Farmers in Japan get returns equal to someone working at Hitachi, but not in America. We tend to undervalue a work ethic that appreciates and supports blue collar employees and working with one's hands. If you are worried about what has happened to the minority population in your universities and colleges, you ought to know that the vast majority of young black men are not pursuing higher education degrees. Rather they are going into the burgeoning proprietary education sector, e.g., Joe's Truck Driving Company, Mark's Tool and Die School, Jim's Plumbing School. One must be really focused on ancillary things if one thinks that learning how to grout a bathtub is unimportant.

Secondly, is this strategic planning process really about niches or is it about vision; or is it about both? Is it about who gets what, rather than what ought we be doing?

Thirdly, are you in the University of Wisconsin System not deep enough, rich enough, diverse enough, self-assured enough to nurture institutional deviation and idiosyncratic behaviors? If you are not, you *are* like Pogo; you are a bigger part of the problem then you are the solution. Is not this system that good, that self-assured to take risks and gamble?

Fourthly, in discussions on curriculum reform, not only is photosynthesis and the need to work against photosynthetic illiteracy critical, but what about civic leadership and what about the notion of *civitas*? Where *is* our common philosophical and ideological edge? Does it only show up when Texans can rob us blind to take care of the savings and loan companies? If anything is missing today in our discussion, it is the notion of *civitas*—civic leadership, civic responsibility, the commonweal, a commitment to us and not just to me.

Finally, let me note that the University of Wisconsin System has given to this country something profoundly powerful that used to be called "the Wisconsin Idea." Where is it? Who's got it? Under what rock does it exist? In whose bushel basket is it being carried around? Where are the great political economists of the past? Economists who can not only count but write; people who can understand things about land economics and who are also concerned about the quality of teacher education. No university in American experience championed more the notion of progressive civic leadership than did the University of Wisconsin. In closing, I challenge you to modify and re-create that Wisconsin Idea.

If this university system will not address the challenge, who will?

NOTES

I express my gratitude to Dr. Robert Brander of the National Park Service, Professors Harvey Jacobs, Richard Barrows and Harold C. Jordahl, Jr., of the University of Wisconsin–Madison, and Professor Robert J. Engelhard of the University of Wisconsin–Stevens Point. Also, Virginia Canter of the Wisconsin Department of Development was most helpful. Of course, they are not responsible for any errors or mistakes in this paper.

BIBLIOGRAPHY

Barrows, R., and Barlowe, R. 1979. "The Farmland Retention Issue: A Midwestern Perspective." Unpublished Master's Thesis, University of Wisconsin–Madison.

Brion, D. J. 1988. "An Essary on LULU, NIMBY, and the Problem of Distributive Justice." *Boston College Environmental Law Review.* 15:3/4.

Cline, F. Y. 1988. "Chernobyl Shakes Reindeer Culture of Lapps." *New York Times.* September 12.

Conklin, H., 1969. *The Nature and Distribution of Farming in New York State.* State Office of Planning Coordination.

Crosson, P. R., and Rosenberg, N. J. 1989. "Strategies for Agriculture." *Scientific American.* 261:128–130.

Department of Agricultural Economics, University of Wisconsin. 1988. *Status of Wisconsin Farming, 1988.*

Dhillon, P., and Derr, D. 1974. "Critical Mass of Agriculture and the Maintenance of Land in Agriculture." *Journal of the Northeastern Agricultural Economics Council.* 3:1.

Grossman, M. 1985. "Exercising Eminent Domain Against Protected Agricultural Lands: Taking a Second Look." *Villanova Law Review.* 30:701–766.

Heberlein, T.A. 1987. "Resource Based Recreation and Tourism." *Resource Policy Perspectives—Stewardship Conference.* Center for Resource Policy Studies, University of Wisconsin.

Jacobs, H., Jordahl, H., and Roberts, J. 1989. "Wisconsin's Twentieth Century Land Policy Legacy." Unpublished Master's Thesis, University of Wisconsin–Madison.

Lapping, M. 1990. "Farmland Acquisition for Public Purposes." *Proceedings, Governor's Conference on Agriculture and the Environment.* New York Department of Agriculture and Markets.

———. 1975. "Land Use Planning: A View from the Courts." *Considerations for Land Use.* Cooperative Extension Service, University of Missouri.

———. 1979. "Underpinnings for An Agricultural Land Retention Strategy." *Journal of Soil and Water Conservation* 34:124–126.

Lapping, M., and Clemenson, H. 1983. "The Tenure Factor in Rural Land Management: A New England Case Study." *Landscape Planning* 10:255-266.

Lapping, M., and Leutwiler, N. 1987. "Agriculture in Conflict: Right-to-Farm Laws and the Peri-urban Milieu for Farming." In *Farming Near Cities*, W. Lockerez, ed. Soil Conservation Society of America.

Lapping, M., Penford, A., and MacPherson, S. 1983. "Right-to-Farm Laws: Do They Resolve Land Use Conflict." *Journal of Soil and Water Conservation.* 38:465–467.

Large, D. 1973. "This Land is Whose Land? Changing Concepts of Land as Property." *Wisconsin Law Review.*

Mandelkar, D., and Cunningham, R. 1985. *Planning and Control of Land Development* Charlottesville, Virginia: Michie.

Novak, P. 1985. "Agriculture and the Natural Resource Base: Status, Problems, and the Future." *Resource Policy Perspectives—Stewardship Conference.* Center for Resource Policy Studies, University of Wisconsin: 95–98.

Prosser, W. 1971. *Law of Torts* West Publishing Company.

Raup, P. 1980. "Competition for Land and the Future of American Agriculture." *The Future of American Agriculture as a Strategic Resource.* S. Batie, and Healy, R. eds. The Conservation Foundation: 42.

Roberts, J. C. 1987. "The Forest Industry and the Small Landowner: Planning and Public Policy Issues." Department of Urban and Regional Planning, University of Wisconsin–Madison, Occasional Papers Series No. 28.

Roberts, J., Tlusty, W., and Jordahl, H. Jr. "Farmers and Their Woodlands: Wisconsin Farmers and Woodland Management." Cooperative Extension Service, University of Wisconsin–Madison.

Roberts, J., Tlusty, W., and Jordahl, H. Jr. 1986. "The Wisconsin Non-Industrial Woodland Owner: A Profile." Cooperative Extension Service, University of Wisconsin–Madison, Occasional Paper Series No. 19.

Ruttan, V. 1980. "Agricultural Research and the Future of American Agriculture." *The Future of American Agriculture As a Strategic Resource* by S. Batie, and Healy, R eds. 138–146.

Scientific American. 1989. "Managing Planet Earth." September, 261:3.

Stier, J., and Jordahl, H. Jr. 1987. "Wisconsin's Forest Resource: Current Status and Prospects for the Future." *Resource Policy Perspectives—Stewardship Conference.* Center for Policy Studies, University of Wisconsin–Madison: 45.

Swader, F. N. 1980. "Soil Productivity and the Future of American Agriculture." *The Future of American Agriculture as a Strategic Resource* by Batie, S., and Healy, R. eds. The Conservation Foundation. 79.

Tlusty, W., and Jordahl, H Jr. 1988. "Politics and Policy in Formulating Integrated Forest Management: The 1985 Wisconsin Managed Forest Law." *Transactions of the 53rd North American Wildlife and Natural Resources Conference.* 46–61.

U.N. World Commission on Environmental and Development (the Brundtland Commission). *Our Common Future.*

Wisconsin Department of Development. 1987. *Summary, An Economic Analysis of Wisconsin's Regions.*

———. 1988. *Interim Report of Governor's Rural Development Coordinating Council.*

———. 1989. *Wisconsin Strategic Planning Council Report No. 1.*

Wisconsin Strategic Development Commission. 1985. *Final Report.*

8

The Viability of Rural Populations and Communities

JAMES J. ZUICHES

In remote rural communities of upstate wherever (New York, Washington, Michigan, Wisconsin, fill in the blank), we can easily observe a struggling agriculture, and poverty-ridden communities. Poor drainage, low soil fertility, and short growing seasons, combined with distance from urban areas, limit dairy farm productivity, and profitability. A university research community recognizes the problems and mobilizes. Local state legislators provide a grant for a revitalization program that draws on both research and extension.

Agronomists focus on forage production problems. They develop new varieties and better management practices. Agricultural engineers analyze soil conditions and develop drainage systems for forage production. Animal scientists and farm management specialists work closely with dairy farmers to improve milking systems, and to computerize herd records and analysis in order to strengthen dairy herd quality. Legislation is passed to include the area in a milk marketing area, and federal funds are used to improve bridges in the area to handle large milk trucks.

Productivity per dairy farm doubles over the fifteen years of the project. Dairy farmers can grow new short season alfalfa varieties on drained fields for their own use. Yields soar because of better herd management, and the profits of dairy farmers increase. A success story of university intervention. Figure 8.1 shows the underlying model of the process and the assumed direction of benefits to the community.

During the same time period, however, 75% of the dairy farmers left the farm and local unemployment continued to rise as a natural resource-based economy, lumbering, collapsed. Only seasonal tourism contributed to the income of the area.

Sociologists or rural development economists were not a part of the university team. No regional planners or community development extension specialists were involved in the effort. As yields increased, no one noticed that the larger community continued to fail.

As I look at this case study, I do not fault the scientists, who in doing their respective jobs well, contributed to the success of the remaining farms. However, I do fault the leadership of research and extension for the blinders we wore, and I do fault my rural social science colleagues for a failure to pursue aggressively the difficult questions associated with agriculture and rural development.

Let us consider another example of change closer to my geographic and disciplinary home. Socio-demographic change is occurring in picturesque Hangman Valley only two miles from Spokane, Washington. The family farms that provide vegetables and other produce for the city are threatened by demographics and development. As the farmers in their 60's and 70's contemplate retirement, their children have gone into careers in electrical engineering, medical technology, and horticulture. In the near future, 3,000 homes and a golf course are planned for this valley. This growth will dramatically change an agriculturally-productive rural setting to a suburban tract with shopping malls, apartments, more traffic and increased service needs. Who will grow the fresh fruits and vegetables for the city of tomorrow?

The strategic planning effort of the University of Wisconsin is an excellent step toward overcoming blinders of the past. Building on an understanding of the past and a sensitivity to the future, the university system is trying to recognize needs and structure programs that can provide solutions for the people of the state. Figure 8.2 is a better representation of the model we must use to address these processes.

"People, not science, transformed U.S. agriculture," wrote James Bonnen of Michigan State University. "Men and women, acting through the institutions which they created, developed scientific knowledge, changed human values and aspirations, modified old institutions, and created new ones as they saw the need, and step by step transformed the productivity and welfare of U.S. farmers" (Mawby, 1986).

EVOLUTION OF LAND GRANT UNIVERSITIES

As we begin a look to the future, it is always useful to reflect a bit on the historical basis of our current effort. The language of the land-grant law signed by President Lincoln on July 2, 1862, remains a profound statement of commitment to education and human resources throughout America:

At least one college where the leading object shall be, without excluding other scientific and classical studies and including military tactics, to teach

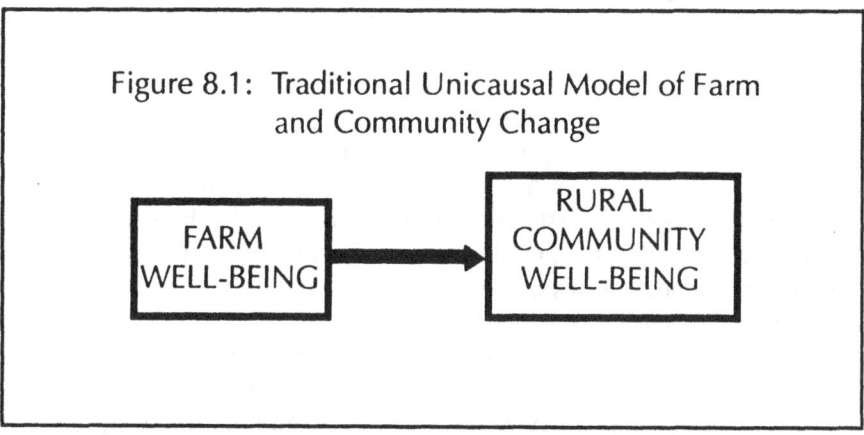

Figure 8.1: Traditional Unicausal Model of Farm and Community Change

FARM WELL-BEING → RURAL COMMUNITY WELL-BEING

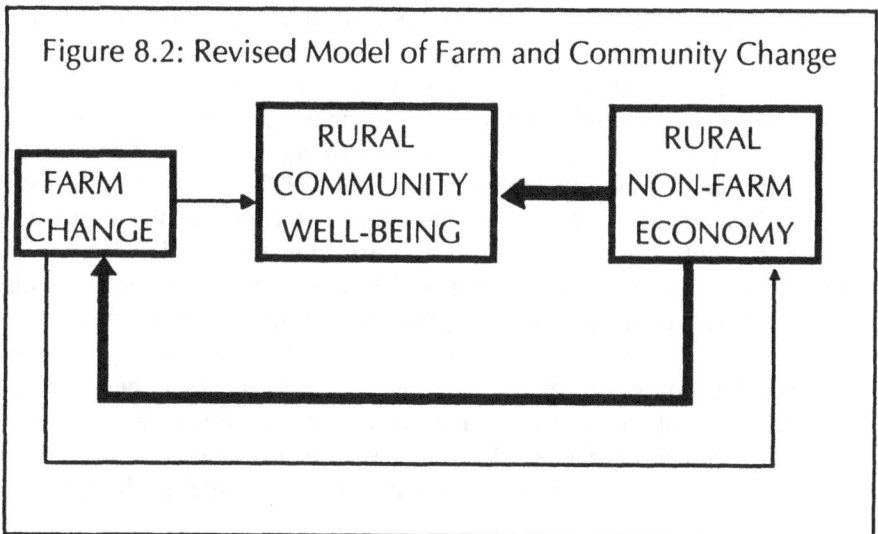

Figure 8.2: Revised Model of Farm and Community Change

FARM CHANGE → RURAL COMMUNITY WELL-BEING ← RURAL NON-FARM ECONOMY

Source: Luloff, A. E. and Louis E. Swanson, eds. *The American Rural Community.* Boulder: Westview Press, 1989.

such branches of learning as are related to agriculture and the mechanical arts . . . in order to promote the liberal and practical education of the industrial classes in the several pursuits and professions in life.

From a modest beginning, the land grant system grew to incorporate and integrate research, instruction, and extension with an orientation to the diffusion of knowledge both within this country and internationally. The history of agricultural research is, likewise, a history of commitment. Yet, tensions and ambiguity exist between the scientific community striving for disciplinary development and advancement and the rural production sector pressing for rapid solution to problems that lessen profitability.

Historically, the scientists and farmers competed for funds, personnel, and research priorities. Agricultural experiment station directors mediate conflicting agendas, reflecting on the one hand the values and norms of the university and its scientific and educational missions and, on the other, needs of the society supporting the university activities. Over the last century, fundamental pressures on land grant system administrators and their research stations have not changed, even though the agricultural research system has grown from an underfunded and understaffed infancy to a maturity able to accommodate an extraordinary diversity in research, instruction and extension programs.

The historical analysis of the early pioneers of agricultural research organizations and administration makes fascinating reading: I recommend the books, *No Other Gods*, by Charles Rosenberg (1966), and *Emergence of Agricultural Sciences*, by Margaret Rossiter (1975). One insight gained from such analysis is the close interaction of experiment station administrators with the agricultural producers, and the associated pressure for immediate research results beneficial to the users. Producer demands were, however, countered with arguments on the value of basic and applied research necessary to understand and then solve client problems. Balancing the autonomy of the scientist and the goals of the research with the needs of the farmers demands a constant communication flow to explain, inform, and even defend institutional activities.

Our work today was presaged by the Hatch Act (1955) which reads, in part:

To conduct original and other researches, investigations, and experiments bearing directly on and contributing to the establishment and maintenance of a permanent and effective agricultural industry of the United States, including researches basic to the problems of agriculture in its broadest aspects and such investigations as have for their purpose the development and improvement of the rural home and rural life and the maximum contribution by agriculture to the welfare of the consumer as may be deemed advisable, having due regard to the varying conditions and needs of respective states.

But it was not always like this. The original Land Grant Act and the original Hatch Act focused on production agriculture and the production problems of plants and farm animals. The Adams Act of 1906 extended basic research on production to all fields. In a recent speech, Russell Mawby (1986), former Executive Director of the W. K. Kellogg Foundation, used this illustration of the problems deans of agriculture faced at the turn of the century because of declining quality of life of farm families and rural communities:

> These deans and directors were responsible for the creation of departments of rural education within the college of agriculture, to address the inadequacies of the one-room school; they created departments of rural sociology, to deal with problems of the broader community; they created departments of home economics, to address the quality of the home and family living circumstances on the farm; they created boys' and girls' club work (now 4-H) to relate formal education to farm living and as a technique to inject innovations.

This effort was typified in the quotation from Liberty Hyde Bailey while still Dean of the College of Agriculture at Cornell University:

> A new social order must be evolved in the open country, and every farmer of the new time must lend a strong hand to produce it. We have been training our youth merely to be better farmers; this, of course, is the first thing to do, but the person is only half trained when this is done. What to do with the school, the church, the rural organizations, the combinations of trade, the highways, the architecture, the library, the beauty of the landscape, the country store, the rousing of the fine community helpfulness to take the place of the old selfish individualism, and a hundred other activities, is enough to fire the imagination and strengthen the arm of any young man or woman. The farmer is to contribute his share to the evolution of an industrial democracy (The Country Life Movement in the United States, 1911).

Such an effort and perspective laid the groundwork for the Purnell Act of 1925. In the Purnell Act, funds were appropriated to pay for the conduct of investigations for

> making experiments bearing directly on the production, manufacture, preparation, use, distribution, and marketing of agricultural products and including such scientific researches as have for their purpose the estab-

lishment and maintenance of a permanent and efficient agricultural industry, and such economic, and sociological investigations as have for their purpose the development and improvement of the rural home and rural life.

SOCIAL SCIENCE RESEARCH IN LAND GRANT CONTEXT

Not too long ago, I reviewed the first year's issues of the journal, *Rural Sociology*. I wanted to assess changes in research focus that might have occurred during the first half century of the journal, so I compared issues of *Rural Sociology* for 1936–37 with those published in 1986–87. In the mid 1930s, articles addressed issues such as economic dependency and need within the community, migration of rural youth out of the community, farm family characteristics, and occupational opportunities of farm families. Institutional changes underway in rural areas (e.g., churches losing their constituents) was a concern. Land use, farm tenure and the loss of farms, living conditions of farm owner and farm laborer families, general relief for rural youth and families, and the problems as well as economic benefits of part-time farming were common research topics. In other words, the array of issues in 1936–37 were not that much different from those facing rural America today.

In the 1980s, net farm income plummeted. Agricultural exports were destabilized. The nation's producers faced severe financial stress as outstanding agricultural debt weighed heavy on many farm businesses. Banks failed, and hundreds of local, agriculturally-related businesses failed as well. Instability was the order of the day. In rural communities there were indicators of stress—the murder of a banker, the suicide of a farmer, or divorce and family disintegration occurring throughout rural America. We had achieved an abundance in productivity, but were losing our profitability and our ability to maintain economic, personal, and societal stability in the face of economic shocks.

What are some of the things we know? What research is needed? What is the role of land grant universities vis-a-vis rural people and communities?

Agricultural and Rural Interdependence

The interdependence of agriculture with rural economic, family, and community conditions has long been a research focus of social scientists. Findings reveal that, over time, large family farm agriculture contributes to higher family incomes, less family poverty, and less unemployment. Corporate-commercial farming shows little relation to socioeconomic well-being. And, although small farm structure may serve to prevent economic circumstances from becoming worse, it is associated with poorer conditions (Reif, 1986, 1987). While the beneficial effects of large family farm structure on socioeconomic conditions of the community may not be surprising, unanticipated, however, was the fact that large family farm agriculture compared favorably with industrial, transportation, and wholesaling employment. However, it is large farm agriculture that appears to have taken the brunt of the 1980s' farm crisis. By implication, declines in this structural pattern can be expected to result in deterioration of economic and social conditions in rural areas.

It is clear that waning incomes in rural areas result in part from long-term agricultural declines. For many rural areas, however, the difficulties lie not only in the adjustment processes of agriculture. Market shrinkage of other natural resource products has hurt rural employment (Miller and Bluestone, 1987). Since 1980, lumber, hardware, furniture, and farm implement businesses show substantial incidences of failure in rural areas (Johnson and Fuguitt, 1987). The mining industry has suffered declines as well (U.S. Bureau of Economic Analysis, 1984).

Rural manufacturing jobs are low-wage, blue-collar, and labor-intensive. They are concentrated in apparel, textiles, and leather goods, all of which have suffered substantial domestic market losses to foreign imports. Thus, market shrinkage in manufacturing goods adds to the economic difficulties of rural areas.

Rural Well-Being

Although the 1950s and 1960s were a period of national metropolitan and suburban population growth, nonmetro growth rates of the 1970s reversed a long-standing trend and exceeded metro rates in all regions except the South (Zuiches and Brown, 1978). But in the mid 1980s, the pattern of population turnaround that promised a renewal for rural America reversed again (Richter, 1985; Engels, 1986; Beale and Fuguitt, 1986). While in the 1970s the proportion of rural counties losing population dropped to less than one-fifth, more recent estimates show that nearly half of all rural counties are again losing population (Drabenstott and others, 1987). Such decline is particularly evident in farming, mining, and other natural resource-dependent counties (Murdock and others, 1987). The most recent surveys of residential preferences show no major resurgence of rural growth. (See Table 8.1)

Fallout from adjustments to the declining rural economy is also clearly evident in an array of personal and family problems in rural U.S. households. Analyses of the rural economic crisis suggest that an increased

Table 8.1

Preferred Residence by Current Residence
Trichotomized

United States: 1972 and 1988

| | | Preferred Residence | | |
Current Residence	Number[a]	City Over 50,000	Other, Less Than 30 Miles From City	Other, More Than 30 Miles From City	TOTAL
1972					
City over 50,000	641	48%	41%	11%	100%
Other, less than 30 miles from city	500	7%	78%	15%	100%
Other, more than 30 miles from city	306	9%	44%	47%	100%
TOTAL	1,447	25%	55%	20%	100%
1988					
City over 50,000	508	60%	29%	11%	100%
Other, less than 30 miles from city	481	10%	78%	12%	100%
Other, more than 30 miles from city	244	9%	39%	53%	100%
TOTAL	1,233	30%	50%	20%	100%

[a] Excludes 34 cases in 1972 and 51 cases in 1988 due to "don't know" or "no answer" responses to either the preference or current residence questions.

Source: Fuguitt, et al, 1989

rate of social, psychological, and emotional problems (e.g., marital discord, spouse abuse, and depression) strike farm producers (Heffernan and Heffernan, 1986; Albrecht and others, 1987) as well as other residents of declining rural communities (Murdock and others, 1987). However, many of these community mental health and related needs are going unattended because of the inadequate level of service programs in rural parts of the United States.

Infrastructure and Information

Information technologies are increasingly important to all types of businesses—from the farm to manufacturing firms. Unfortunately, rural America has a greater proportion of party lines, but less digital switching, less trunkline capacity, poorer line quality, less optic fiber, fewer computers, and fewer of the other advanced information technologies (Dillman, 1985). Overcoming these problems and finding ways to encourage the adoption of these technologies by rural households and businesses represent a possible means for revitalizing rural America.

Intergovernmental Relations

During the 1960s and 1970s, state and federal involvement in the affairs of local government grew. By 1977, intergovernmental transfers represented 43% of total revenues for rural localities compared with 34% in 1962 (U.S. Bureau of the Census, 1965, 1980).

By 1990, the range of federal actions affecting rural areas—and the policy latitude of state and local governments—have changed substantially. Federal funding has been withdrawn or reduced for many grant-in-aid programs, block grant programs, and revenue sharing. In part, these changes represent a philosophical retrenchment by the federal government and, in part, a policy to reduce federal deficits while maintaining defense and major entitlement programs. This is a challenging environment in which state and local governments find it difficult to replace federal funds with local revenues (Stinson, 1986).

Global Economy

The importance of international trade discussed by Tutwiler has increased the stake of rural areas in macroeconomic trade policies. Many jobs in low-wage rural manufacturing are vulnerable to foreign competition. Compared with the federal government, states and localities are severely limited in policy responses to deal with industrial restructuring and trade (Brown and Deavers, 1987).

Other Socioeconomic
and Demographic Trends

Several other trends discussed by Heltsley and other authors should be mentioned to complete the picture. They include:

1. The urbanization/suburbanization of rural land continues unabated.
2. We have an aging society, with many seniors selecting rural locations for retirement. For example, Wisconsin ranked 14 in the number

of residents over 65; ranked 15 in percent over 65; and had a 10.7% change from 1980 to 1986 in the size of the aged population.

3. There is more widespread recognition that family needs, especially youth-at-risk, demand changes in social service delivery in rural areas.

4. As a result of small families, dual incomes, and delayed births to baby boomers, productivity is likely to increase for families.

5. With more women in the workplace, issues of income equity and childcare infrastructure have been added to agendas of government bodies and voluntary organizations in rural communities.

6. The workforce transformation—a greater percentage of minorities and women, and individuals with different educational back-grounds—means a significant shift in the quality of the labor force.

7. Continued growth of knowledge-based industries—collecting, ana-lyzing, synthesizing, storing, providing and retrieving information for decision-making, will require expanded training for the new labor force.

8. Society is experiencing value shifts to a more conservative position, (e.g., protection of investments, health benefits, right to life) yet non-traditional values, such as food safety and quality of envi-ronment, are concurrently reflecting other needs of society.

Rural Wisconsin is not immune to these trends. Rural Wisconsin is also in trouble. Economically, Wisconsin ranked 46 among the 50 states in both nonmetropolitan and metropolitan new enterprise job growth, 42 of 50 states in the number of new companies per 10,000 workers. It ranked 29 in the percent of fast growth companies.

Yet, Wisconsin has tremendous human resource potential. It ranked 9th in the percent completing high school and 14th in the nation in the level of college attainment. Science and engineering were important pursuits by its citizens; it ranked 14th in the percent of students in science and engineering and 25th in percent of the labor force in these two areas. University expenditures for R & D in dollars was 8th in the country, while federal expenditures for R & D ($) was 41st in the nation.

RESEARCH NEEDS

Farmers, citizens, and community leaders, as well as state and local officials, are asking questions about agriculture and rural viability. They raise questions for which answers will require the research anticipated here:

1. How many more farmers must leave farming?

2. How can we maintain and develop new jobs in rural America?
3. How can we best assist individuals and families who must leave farming, nonfarm jobs, and businesses?
4. What types of communities are most or least likely to remain viable in the future?
5. What are the stresses of farm and business failures on families and communities? And what are the best ways to help?
6. What do the new agricultural technologies coming down the road—biotechnology, robots, computers, telecommunications, and the like—mean for the future structure of agriculture and rural communities?
7. What are some profitable new enterprises using farming resources?
8. Who is going to provide the leadership in rural America in the twenty-first century?
9. How can we provide services, especially in areas that experience abnormally high out-migration?
10. How can we maintain a sense of community in rural areas with all these changes?

The same types of questions are being addressed by other groups like regional governors' associations; the Council of State Governments; the Association of Counties, Towns and Townships; and the League of Cities. (See Table 8.2).

Can we answer these questions? I would contend that the answer is yes.

We have a system of agricultural research, instruction, and extension that gives us a comparative advantage in responding to these sources of change. We have an institutionalized mechanism, an infrastructure, for responding. Colleges of agriculture and home economics are shifting their courses of instruction to provide the kind of education students need. Cooperative Extension services around the country are responding to the stress experienced by farm families, providing financial program assistance and sources of support, referral services, and the information that one needs to make decisions in a crisis. Research programs are being developed to buttress the instructional and extension efforts in these arenas.

However, a major difficulty is faced in developing policies and programs to aid farmers and other rural people as they attempt to adjust to the current economic problems in rural areas. The difficulty is that we know so little regarding changes in the agricultural and natural resource economies and their interdependence with community viability and family well-being. We do not have adequate answers to many basic

Table 8.2

Impacts of Changes in Social Structures, Devolution, and Revitalization on Information Decision Making in Rural Governments.

New Development	*New Information Implications*
Restructuring Movement by societies to new population, economic, spatial and public-private sector dynamics, patterns, and linkages	Need for intelligence and soft data or early warning information on incipient changes of dynamics, patterns, and linkages Need for hard data to identify and confirm trends, to monitor change, and to predict alternative socioeconomic futures and market tendencies
Devolution Transfer or shift of fiscal and functional responsibilities from central/federal to local (rural) governments	Loss or disruption of rural-area data collection in censuses and surveys, and rural area statistical series Loss or disruption of rural-area socio-economic and environmental studies, and professional/technical expertise and consultation Need for data to evaluate existing programs (for continuation), and as basis for creating new or replacement programs
Revitalization Rapid expansion or diversification of economic base, population profile, housing market, social services support systems, physical infrastructure, land use activities, etc.	New data needed on available and required land, infrastructure, social services, and (local) finances to accomodate or promote revitalization New data or intelligence needed on private sector intentions, development spin-offs, and citizen preferences on rate, scale, and direction of revitalization

Source: Wellar, B. S. 1984. Cited in L. E. Garkovich, "Local Organizations and Leadership in Community Development."

questions being asked by rural leaders who are accustomed to finding answers in the land grant university.

To answer these questions attention must be given to the new research initiative proposed by rural social sciences. The goal would be to accomplish at least the following objectives:

1. Develop a more complete and accurate picture of the emerging conditions in rural areas of each state.
2. Determine the directions and persistence of trends in agriculture, rural families, rural communities, and the rural economy.
3. Assess the impacts of changes in the structure of agriculture on local economies.
4. Identify farm and off-farm community economic development opportunities.
5. Develop more effective systems of stress management for farm and rural families.
6. Develop and evaluate methods of assisting displaced farm and rural families.
7. Evaluate alternative systems for providing public services in rural areas.
8. Assess the impacts of national farm and nonfarm policy on rural communities.

THE ROLE OF THE UNIVERSITY

What is the role of higher education in rural development? In human development? In the interdependence between natural resource-based economies and the family and community?

First, the land grant university provides access to knowledge and technology needed to address these issues. This research base must be university-wide, not simply in colleges of agriculture, or home economics, or natural resources and environment.

Second, the research must address the problems of society and in its applications directly meet the needs of the state. I have argued consistently in the research programs that I administer for a balance between basic, long-term scientific understanding of issues and applied, problem-solving near-term scientific analysis. Both must exist in a land grant university.

Third, the knowledge base must be delivered to the people. If the knowledge is technology, then a transfer mechanism is needed; if the knowledge is non-technological, then an instructional system is needed. If Cooperative Extension did not exist, we would have to invent it again to help solve the issues of today.

Fourth, we must educate and train the citizens of the state, not just during the 4–5 years of residence in a dormitory, but over a life-time. The social, economic and structural changes I have been describing will not be solved with a one-time infusion of knowledge. We know that in the physical and biological sciences; it is true in the social and economic sciences as well.

The underlying issues are capacity building in the individual, family, community, firm, organization, and governmental body to enable these entities to be more successful in their day-to-day interaction with social change. The primary value of the university to the leadership of a community or state is to provide access to knowledge.

One mechanism is to establish an Institute for Community and Rural Studies. Cornell University, the University of Massachusetts, and the University of Georgia are creating such institutes or centers. These organizational entities are required:

1. as a point of focus for efforts to serve a network of constituents and scholars as a central focus of communication and collaboration;
2. as a point of entry for external constituents approaching the university for assistance, research, or educational programs;
3. as a central exchange for research and extension efforts associated with site-specific programs;
4. as the source of information and implications on trends, shifts, consequences. Data needs are evident in all areas as shown in Tables 8.3 and 8.4;
5. as a core focus for interdisciplinary, cross departmental research and extension teams.

These efforts extend the concept of biotechnology centers, centers for advanced technologies, or research and technology parks to include data collection and policy analysis that is often ignored in the decisions to locate such centers. These institutes need to be mission-oriented. But even mission-oriented efforts require an investment in multiple activities: disciplinary or subject-matter and problem-solving research, technology development and demonstration programs, adaptive research, mainte-nance research, and knowledge transfer to the users in the class or in the state. Social science alone cannot achieve sustainable community systems. But in alliance with local and state organizations, business and labor, government and education, we can contribute to the viability of rural people and areas.

The genius of the system is that over time it creates intimate contact among problems of the people, researchers, extension agents, policy-

Table 8.3

The Concerns of Scholars and Local Leaders are Focused on Issues:
  Rural Families - Change and Continuity

Organizational Well-Being	*Family Well-Being*	*Community Infrastructure and Well-Being*
Leadership	Domestic violence	Rural schools quality
Rural Ministry	Child well-being	Rural health care availability
Conflict resolution	Single-parent families	Adult educational needs
Farmer worker employment	Helping rural youth	SSAN benefits, rehab
Successful family farms	Dating violence	services, and
Managing volunteers	Parenting	independent living
Revitalizing a town	Weight control	Out-reach to financially
Business development	Teen sexuality	stressed families
Sustainable communities	Drunk cowboy syndrome	Rural social programs
Rural networks	Child abuse and neglect	
Financial management	Teen pregnancy	
Financing retirement	Adolescent family	
Media and rural families	Stress	

Source: Kansas State University, Program Announcement, 11/15-17/89.

Table 8.4
  Implications for Colleges and Universities of the Changing U.S. Economy

Traditional Economy	*New Economy*	*Implications for Universities*
Slow-moving technology	Rapid technical change	Increased research Technology transfer
Distinct technical fields	Merging technical fields	Interdisciplinary programs or centers
Little foreign competition	Strong foreign competition	Knowledge of new competition
Focus on domestic markets	Focus on global markets	Knowledge of new cultures, languages
Mass-produced products for mass markets	Complex products for sophisticated consumers	Technical aid for business
Growth in volume of products sold	Growth in value added to products sold	New products and more flexible processes
Human resources as a cofactor of production	Human resources as a competitive edge	Stronger educational system
Slow-changing skill requirements	Rapidly changing skill requirements	Lifelong learning, extension programs
Employment growth in Fortune 500	Employment growth in new/small firms	Support for entrepreneurship
Economic growth through smokestack chasing	Growth through new business development	Commercialization of new technologies

Source: Treadway, Douglas M. "Higher Education and Rural Development: The Next Frontier." *Northwest Report* October, 1988.

makers, and federal, state, and local government units. We created this system, but we can lose it as well if it is not renewed.

BIBLIOGRAPHY

Albrecht, Don E., Steve H. Murdock, Rita R. Hamm, and Kathy Schiflett. 1987. "Farm Crisis: Impact on Producers and Rural Communities in Texas." College Station, Texas: Texas Agricultural Experiment Station, Department of Rural Sociology Technical Report 87-5.

Bailey, L.H. *The County Life Movement in the United States.* New York: The MacMillan Company, 1911.

Beale, Calvin L. and Glenn V. Fuguitt. 1986. "Metropolitan and Nonmetropolitan Population Growth in the United States Since 1980." Pp. 46–51 in *New Dimensions in Rural Policy: Building Upon Our Heritage.* R.C. Wimberley, D. Jahr, and J.W. Johnson, eds. Washington, D.C.: Joint Economic Committee, Congress of the United States.

Brown, David L. and Kenneth Deavers. 1987. "Rural Change and the Rural Economic Policy Agenda for the 1980's." Chapter 1 in *Rural Economic Development in the 1980s: Preparing for the Future.* AGES-870724. Washington, D.C.: Economic Research Service, U.S. Department of Agriculture.

Corporation for Enterprise Development with Mt. Auburn Associates. *The 1989 Development Report Card for the States.* March 1989.

Engels, Richard A. 1986. "The Metropolitan/Nonmetropolitan Population Redistribution 1972–1988." Paper presented at the annual meeting of the Rural Sociological Society, Seattle, Washington, August 1989.

Fuguitt, Glenn V. et al. "Residential Preferences and Population Redistribution 1972-1988." Paper presented at the annual meeting of the Rural Sociological Society, Seattle, Washington, August 1989.

Garkovich, Lorraine E. "Local Organizations and Leadership in Community Development." Chapter 10 in *Local Organizations and Leadership.*

Heffernan, William D. and Judith B. Heffernan. 1986. "The Farm Crisis and the Rural Community." Pp. 273–280 in *New Dimensions in Rural Policy: Building Upon Our Heritage.* R. C. Wimberley, D. Jahr, and J.W. Johnson, eds. Washington, D.C.: Joint Economic Committee, Congress of the United States.

Johansen, Harley and Glenn Fuguitt. 1987. "American Villages in the 1980s: Effects of Development Efforts, Location, and Agriculture Economy on Business Trends." Presentation at meetings of the AAG, Portland, Oregon.

Mawby, R.G. "Evolution of the Land-Grant System." *Issues Facing Agriculture and Implications for Land-Grant Colleges of Agriculture.* Proceedings of a workshop October 9–10, 1985, for Deans and Directors North Central Region, Oakbrook, Ill. Farm Foundation, 1986.

Miller, James P. and Herman Bluestone. 1987. "Patterns of Employment Change in the Nonmetropolitan Service Sector, 1969–84." Paper presented at the meetings of the Southern Regional Science Association, Atlanta, Georgia.

Murdock, Steve H., Don E. Albrecht, Rita R. Hamm, F. Larry Leistritz, and Arien C. Leholm. 1986. "The Farm Crisis in the Great Plains: Implications for Theory and Policy Development." *Rural Sociology* 51:406–435.

Reif, Linda L. 1987. "Farm Structure, Industry Structure, and Socioeconomic Conditions in the United States." *Rural Sociology* 52 (Winter):462–482.

Richter, Kerry. 1985. "Nonmetropolitan Growth in the Late 1970's: The End of the Turnaround?" *Demography* 22 (2):245–263.

Rosenberg, C.E. *No Other Gods.* Baltimore: Johns Hopkins University Press, 1976.

Rossiter, M. *The Emergence of Agricultural Sciences—Justus Liebig and the Americans, 1840–1880.* New Haven: Yale University Press, 1975.

Stinson, Thomas. 1986. *Governing the Heartland: Can Rural Communities Survive the Farm Crisis?* Washington, D.C.: U.S. Senate Committee on Governmental Affairs.

Swanson, Louis E. "Rethinking Assumptions about Farm and Community Change." Chapter 3 in A.E. Luloff and Louis E. Swanson (eds.). *The American Rural Community* Boulder, Colorado: Westview Press, 1989.

Treadway, Douglas M. "Higher Education and Rural Development: The Next Frontier" Pp. 8 in *Northwest Report* October 1988.

U.S. Bureau of the Census. 1986. *Money Income and Poverty Status of Families and Persons in the United States: 1985.* Current Population Reports, Series P-60, No. 154. Washington, D.C.: U.S. Government Printing Office.

U.S. Bureau of Economic Analysis. 1984. Unpublished data, Washington, D.C.: U.S. Department of Commerce.

Wellar, B.S. 1984. (cited in L.E. Garkovich "Local Organizations and Leadership in Community Development" Chapter 10.). Information for decision making by rural public authorities in *Local Leadership and Rural Development: Implications for Research and Extension.* Organization for Economic Cooperation and Development. Proceedings from the Conference on Local Leadership and Rural Development. Colonial Williamsburg, VA., April 19–30

Zuiches, James J. and David L. Brown. 1978. "The Changing Character of the Nonmetropolitan Population 1950–1975." Chapter 4 in T.R. Ford, ed. *Rural USA: Persistence and Change.* Ames, Iowa: Iowa State University Press.

NOTES

Much of the section titled "Social Science Research in Land Grant Context" is quoted from or derived from "Agriculture and Rural Viability," Experiment Station Committee on Organization and Policy, Cooperative State Research Service, USDA, 1988, 88-3, which was prepared by the Task Force on Agriculture and Community Viability, co-chaired by James J. Zuiches and Richard J. Sauer.

9

It's in Someone Else's Backyard, So Why Worry? Rural Youth and Rural Families

MARY E. HELTSLEY

As a culture, we are often self-centered and shortsighted about our concerns, issues, and priorities. Even though we do long range planning, our emphasis is usually immediate rather than anticipatory. However, the strategic planning process encourages us to be anticipatory and proactive. While this is risky, it is also exciting—one is more likely to win some and lose some rather than just play the surefooted game. Moreover, when we take risks, the gains can be noteworthy. Issues need not be considered problems, but instead can be seen as attainable challenges.

An example that illustrates a lack of concern and forethought is the AIDS crisis in our country. Although the AIDS epidemic has been raging in Africa for a number of years, we have not significantly addressed this issue in the United States. Even after we had evidence in the United States that AIDS was a problem among homosexuals and intravenous drug users, it was not our problem since "it's in someone else's backyard, so why worry?" I remember that a few years ago when I moved from the East Coast to the Midwest I found it difficult to believe that major metropolitan newspapers rarely made reference to the AIDS epidemic. In one of my "say a few words" speeches as a new dean, I strongly stated that AIDS was an issue that should concern social science and social service professionals. I felt like an alien—as if I had introduced a topic that was out of context. As administrators, however, it is important to support researchers who redirect their energies from related areas of work to emerging issues, even if it is risky.

Social change has taken place very rapidly during our lifetime. We have lived through the Pepsi generation, the babyboomers, and the yuppie generation. We are headed for an even more exaggerated form of the "sandwich generation." In 2000, the U.S. population will have a preponderance of the young and the old, most of which will also be members of minority groups. The squeeze will be on those caught in the middle—those who are now the yuppies. As a society, however, we often pay little attention to issues such as this that seem to concern others and not us. We do little to anticipate and avert the coming crises of the world around us.

What does all of this have to do with the price of milk on a Wisconsin farm? Youth, family, community and elderly problems occur everywhere. Although some youth or family problems appear specific to either rural or to urban areas, most problems affect both areas in one of two ways: 1) the problem, issue or event occurs in urban areas, then appears at a later time in rural areas, or 2) although the problem, issue or event exists in both rural and urban areas, it manifests itself in a different form in each place.

Two examples illustrate these processes. First, the AIDS crisis, for example, is an issue appearing first in urban and then rural areas. No reported deaths from AIDS had occurred in Minnesota in 1987, however AIDS deaths were reported daily in the newspapers on the East and West Coast. Not long after that the Twin Cities newspapers reported first one death, then two deaths, and now the numbers have accumulated into the hundreds. In 1989 we find that AIDS is in our backyard and therefore, it is time to be concerned. We are noted for being socially conscious people, yet, we also fail to recognize danger signals at times.

Second, crime illustrates a problem that manifests itself in different forms in urban and rural areas. Crime rates remain consistently higher in large cities than in non-metropolitan areas. For most crimes, the rates for non-metropolitan areas mirror those of urban areas; however, there is little evidence that crimes begin in large cities and then spread to rural areas. Rather, large cities and small towns are affected almost simultaneously by national crime trends (McGranahan, 1986:2–8). Violent crimes have generally risen more quickly in metropolitan areas, and property crimes rose more quickly in non-metropolitan areas. Declines in crime rates in 1980s in the non-metropolitan areas may reflect rural residents' and businesses' awareness that they, like their city cousins, are not immune to crime. However, rates for burglary without forcible entry are as high in non-metropolitan as metropolitan areas, perhaps reflecting the trusting nature of many rural residents.

According to the U.S. Bureau of the Census classification, *urban* refers to a place that has 2,500 or more people per square mile and *rural* refers to all the others with lower population density. Another frequently-used census dichotomy classifies areas into *metropolitan*—counties that contain at least one city or two adjacent cities of 50,000 or more population, and *non-metropolitan*—all other areas. In the first classification, the small town, a characteristic of the Midwest, could be listed as *urban*, even though it may differ little from the rural areas. In contrast, the *metropolitan/non-metropolitan* classification may be more useful in differentiating lifestyles and attitudes in the rural and urban Midwest.

Dividing the population into rural and urban segments on youth and family issues may be shortsighted. However, it is customary for academic institutions to justify their existence in part by drawing their boundaries and establishing their priorities. But, no matter where one lives, problems emerge for youth, families, the elderly, and communities. Administrators in colleges of agriculture and natural resources must address these problems through research, long-range planning, and risk taking wherever they occur—the city or the country—because youth are integrated in a common bond of being young and wanting to experience the world. The first step, however, is to identify the problems at hand.

RURAL YOUTH

Youth problems are now in our backyard—and it is time to worry. Teen pregnancy, teen prostitution, and school drop out rates are just a few of the many youth problems facing America today. Teenage pregnancy suddenly became a rural issue when one of the rural communities in Iowa was identified in the early 1980s as having one of the highest rates of teenage pregnancy in the country. Such promiscuity, in an "isolated town in Iowa" may be reported by newscasters lacking other interesting stories; however, it may indicate that the rural family value structure is under siege. The statistics reveal that teenage pregnancy is not a phenomenon of an isolated town in Iowa, but is an issue in much of rural America.

Another youth issue, teenage prostitution, has usually been considered an urban issue. However, a sizeable number of the teenage prostitutes found in cities, located in predominately rural states, are actually runaway rural youth. One way in which rural youth can make money in urban areas, other than at fast food outlets, is to perform sex with adults. Rural youth may lack experience in dealing with street life, and thus, fall prey to pimps and others who provide attention and make promises of a better life.

The preceding examples of youth issues both lead to another problem: AIDS. This trend is very alarming to medical and social service professionals. New studies indicate that the infection rate is far higher for teens than for adults. Since many teenagers have multiple partners and rarely use condoms, the spread of the infection is multiplied. The Centers for Disease Control (CDC) in Atlanta estimates that 1 to 1.5 million individuals in the U.S. are infected with AIDS, a rate of approximately 5 out of 1000 (Kolata, 1989). In New York City, 1 in 1000 babies born to 15 year old mothers since 1987 have the AIDS virus antibodies, and 1 in 100 babies born to 19 year olds have the antibodies to the virus.

The CDC study found that even in parts of the country relatively unaffected by AIDS, 3 in 1000 fifteen and sixteen year old mothers were infected; for 21 year olds, the rate was triple this figure. Since the average time from infection to development of AIDS is 10 years, many adults have contacted the disease in adolescence.

If youth are our greatest resource, then what are we doing to reclaim them? The National Commission on Excellence in Education issued the warning that "Our Nation is at Risk." Historically, Americans have made steady progress in completing school, from 13% in 1900 to 76% in 1987. Overall, the number of dropouts has declined from the early 1970's and the gap between blacks and whites has narrowed significantly. Yet, experts warn of a two-tier society composed of a highly educated and prosperous white sector outweighed by a poorly educated and economically deprived group of whites, immigrants, and non-whites. As one analyst wrote:

> Youth, especially minority youth, have little confidence in the deferred gratification that education promises, or the mythical guarantee that a diploma translates into equitable career/employment opportunities (Kunisawa, 1988:61).

Clearly, youth literacy and education are major concerns for the beginning of the next century.

CHANGES AND PROBLEMS FOR RURAL FAMILIES

Perhaps some of the problems incurred by our youth stem from problems created by the family network. For example, one concern in the family setting is "latchkey" children who spend about two hours per day alone after school. An increase in single-parent and dual career families has been instrumental in creating an estimated 2 to 6 million latchkey children. These children have been found more likely to abuse alcohol, tobacco, or marijuana. In one study among eighth graders this finding held whether from affluent or poor families (*Minneapolis Star Tribune*, September 1989).

Domestic Violence

Another familial problem that may spark problems in our youth is domestic violence. Although domestic abuse has always existed in our society, the acknowledgment that it occurs in all age, income, and social groups is a fairly new phenomenon. With the stress placed on rural families during the family farm financial crises of the 1980s, domestic abuse became such an issue that the significance of the extension agent's information about production, tillage, conservation and integrated pest management paled by comparison.

Although the frequency of domestic violence is difficult to compute given the numerous unreported cases, the statistics on reported cases are alarming:

- over one half of the female homicides are the result of domestic violence by the husband or boyfriend;
- more than 1.8 million women in the U.S. are reported battered, 60,000 in Minnesota alone; and
- more than 1 out of 4 American women can expect to be battered sometime during their lives.

While these figures focus on the plight of women, child abuse is equally prevalent. However, elder abuse appears to occur frequently and may reach major proportions as the population ages. Much of the elder abuse occurs because of the long term demands made on younger and over-stressed family members who must deal with the ravages to the body and mind like Alzheimer's, and with debilitating diseases and senility. Undertrained and often uncaring attendants in overcrowded long-term health facilities also contribute to the number of elderly who are abused. As more rural females enter the labor market, a greater portion of the responsibilities for caring for the aged will shift to private long-term health care.

Working Women

Another change facing the rural as well as urban family in the U.S. is women's labor force participation. Social and economic forces have propelled women into the paid work force. Today, the 56.2 million American women in the civilian work force make up 45% of the work force compared with 38% in 1970. By 2000, expectations are that women will represent almost one-half of the labor force (Cowan, 1989). Control over family size, divorce and single parenthood, and young families who need more than one paycheck are all factors pushing women into the work place. Only about 10% of women enter the workforce in search of fulfillment (Cowan, 1989).

Over the last 20 years, the American workplace has been greatly changed by the entry of women into more influential jobs. Women have made dramatic strides in entering traditionally male occupations. However, less economic progress has been made in women's earnings in relation to those of men. Today, women earn on the average 70 cents for every dollar earned by men; this is compared to 62 cents per dollar in the 1960s. In addition, women are almost totally absent from top management jobs and the better-paying, unionized ranks of some professions (Cowan, 1989).

Women's progress in the working world is visible to the public, but the controversies over the nature of this progress abound. Both women and men are questioning the price they have paid for this change and the implications for their private lives. They are torn by the conflicting demands of their jobs and the desire to have more time with the family. Approximately one-half of the women in one study reported that they had had to sacrifice too much for their gains (Cowan, 1989). Children, personal time, and family life were cited as the primary casualties for that sacrifice.

Child Care

Child care is another major concern of urban and rural families. Accompanying the overall change in women's participation in the work force, has been an increase in the percentage of working mothers with children under 18 years and; from 1980 to 1988 this figure increased from 53 to 69%. Although reports indicate that about one-half of all mothers with children under three are not in the labor force, mothers with preschoolers represent the fastest growing group. For children from families with working mothers, almost 25% are placed in group child care centers, 22% go to a caretaker's home, 15% are cared for by relatives, 16% are at home with their fathers, and the remainder have caretakers who came to the family's home.

Housing

One of the reasons that many young couples, rural or urban, find that both must work is to afford the increasing costs of housing. Home ownership peaked in the U.S. at about 66% in 1980. The most dramatic change has been among those in the younger age group. In 1973, 23% of those under the age of 25 owned a home; by 1988, the rate was 15% (*New York Times*, 1989). While home ownership for other age groups has also declined, the differences are less.

Rising real estate costs and higher interest rates are often cited as the reasons for declining home ownership. Housing affordability is determined by several factors, including tax policy, income growth, and the housing market. In addition, increasing down payment requirements affect the ability of young people to purchase homes. The average down payment has risen more than 50% in this decade, from $8600 in 1981 to $13,000 in 1988 (*New York Times*, 1989). Another major hurdle, high interest rates, can be illustrated by the cost of a 30-year mortgage for $100,000. The prevailing rate in the 1960s, 6%, resulted in total interest costs of $115,481, and a monthly house payment of $600, whereas at today's 10.5%, the interest would be $229,306—nearly double the 6% cost—and the monthly payment $915.

At the other end, those who already own homes in some parts of the country are finding that, because of over-building, the value of their homes has fallen sharply—as much as $20,000 to $50,000. Long-term depression of the housing market in some parts of the country, including agriculturally dependent rural areas, makes mobility and job changes difficult.

The inability to buy a home is leading to shortages of adequate housing for low income families. Although the housing shortage and homelessness may seem to be more typically urban problems, the problems are haunting to anyone facing the issue of inadequate, scarce, and low quality housing. The purpose of a recent protest march in Washington, under the banner of "Housing Now!" was to influence the present administration regarding the shortage of decent, affordable housing everywhere. Some rural areas are spared the necessity of coping with large numbers of the homeless because of lack of industry, and minority groups which often are the first to be affected in metropolitan areas. In the Upper Midwest, it may be the harshness of the winter weather rather than our planned response to the need that keeps the numbers low.

What is the future of housing in rural communities as off-farm employment increases and farms decrease? One prognosis is that Levitt's view of planned cities (Levittown) may reemerge with a different emphasis. Rather than satellite communities surrounding metropolitan areas, Levitt envisions new cities of approximately 50 to 250 thousand, that would be "primary employment towns." These would be balanced centers of industries, varied housing, schools and cultural amenities located in what are now rural areas.

Although Levitt was often scoffed at in the past, the idea of proximity is now superfluous. Today, many are asking "why not bring industry to places where living is easier?" With telecommunications, fax machines, computers, and express mail, rural communities may become dispersed, mid-sized urban neighborhoods.

Drugs and Crime

Problems for families are also problems for rural communities. In the late 1980s, for example, drug arrests in rural counties have increased more than twice as fast as in cities. Drugs like crack, thought to be confined to the toughest urban neighborhoods, have been a source of violent crimes in rural areas. Drugs, suicide, homicides, and violent crimes have become intertwined as causes and consequences, often making it difficult to separate one from the other.

However, it is not only drugs that lead to crime. The variable most directly linked to non-metropolitan violent crime is broken families. Single parent families and poverty are interrelated and the two variables create an interesting dichotomy with respect to property crime: the propensity to steal is higher when children come from divided families, but, because many divided families live in high poverty areas, opportunities for property crimes are low.

Health Care

Another problem facing the families and communities in the U.S. is the issue of health care. The problems include coping with rising health costs, chronic diseases of aging individuals, large numbers of individuals needing treatment, and the challenge of providing adequate care. Rural communities will be faced with the additional problem of availability of adequate health care facilities locally. U.S. health care costs have escalated 17% during the last three years, with the greatest out-of-pocket expenses going for health plan premiums. These increases result from rapid medical inflation, use of expensive treatments and diagnoses, and governmental limitations on payments in its programs. One of the greatest costs, however, is paying for the uninsured.

More than 30 million Americans, many of whom hold jobs, have no health insurance. Many businesses find it difficult to finance insurance to cover escalating health costs. The hardest hit employers are small businesses, those with fewer than 500 employers, who account for 53% of the American workforce. These small businesses, many of which are service oriented, have created two of every three new jobs in the past six years (Kramon, 1989). Yet, staggering insurance rate increases of 20 to 100% have strained these small businesses trying to provide health coverage for their employees.

Hospitals, particularly those in the inner cities, are facing similar economic challenges. It is not unusual for public hospitals today to lose millions of dollars a year in caring for indigent patients. Many public hospitals that care for the poor have growing numbers of AIDS patients, many of which are uninsured. Although rural hospitals have not felt the magnitude of the AIDS problem, even limited numbers of cases can cause a severe burden. On the other hand, a more common problem for small rural hospitals is the struggle with net patient revenue margins to provide basic equipment and services. Indigent rural patients, often aged, create severe financial problems for small rural hospitals.

Rural hospitals have other problems, too, as population shifts have reduced the market for health care services. Since lack of mobility restricts access to medical care, the reduction in health care services most adversely affects older rural residents. A differential payment system further discriminates against rural hospitals whose major clientele is often aged adults relying on Medicare payments.

Furthermore, doctors and other medical practitioners are becoming more specialized. The patient base in most rural communities will not support an array of specialists. In addition, doctors are often reluctant to locate in rural areas that have limited medical support and other professional opportunities. Medical malpractice insurance premiums, medicare reimbursements, and financial rewards in specialized fields make it harder to attract the young doctor to primary care. Attracting young physicians is critical. Increasingly alarming to many rural areas is the age of rural physicians. Approximately 20% are more than 65 years of age and another 20% are between the ages of 55 and 65 (Straub and Walzer, 1988). If this trend continues, health care in rural areas could become non-existent.

RURAL ELDERLY

In the next 40 years, the elderly population in the U.S. is expected to double. Currently, about 10% of the population is 65 years of age and over; but residents who are 65 years or more make up 13% of the 1980 non-metropolitan population. As the babyboomers mature, obviously the proportion of the people age 65 and over will increase. The most rapid increase is expected between the years 2010 and 2030; by 2020, 20% of the U.S. population will be over 65 years of age (Glasgow and Beale, 1985; Rieken, 1989). Farm people often move into a nearby village or town when they retire, so the proportion of the small town population over 65 may be even higher.

Older rural people are more likely to be poor than the urban elderly. "Half of all older people in America with poverty level incomes live in rural areas and small towns. . . ." (Glasgow and Beale, 1985:24). The rural elderly are more likely to be poor (21%) than the rural nonelderly (13%). Although growth in social security, Medicare, and other government programs has tended to equalize living conditions for older Americans, the most vulnerable to poverty are likely to be the frail, minority, widowed, and aging females.

For some older residents in rural and small town America, the difficulty of entering or reentering the labor market is a factor in the low income levels. Older workers learn this lesson as they take early retirement, government pensions, or assume that their experience will transfer into the private sector. Others are casualties of corporate reorganization and/ or downsizing. Older employees are often considered dispensable and are encouraged to retire. For many companies, older employees are not as readily sought as younger applicants. While older employees may find the labor shortage in the service industries to their benefit, they are generally offered entry-level jobs, far from their position and pay expectations.

Health care is not only a primary concern in rural areas, but also to the elderly. In 1989, Congress expanded health-care benefits for older Americans, and it is the first time that the cost of Medicare coverage was based on an individual's income. The Congressional bill covers a smorgasbord of needed medical benefits: capping yearly out-of-pocket expenses for Medicare recipients and eventually covering, after the deductible, 80% of prescription costs. The program provides 150 days of skilled nursing care; 38 days of home health care, plus 80 days of paid care to relieve family members; and eliminates the need for exhausting assets prior to obtaining long-term nursing care under Medicaid (*Newsweek*, 1989).

A major problem remains: how to pay for this program? Since much of the costs would be borne by the aged themselves, higher-income elderly will be hit the hardest. While many in Congress see the need to retain program provisions, the outcry from higher income elderly is so strong that the legislation will likely be repealed. A new initiative headed for Congressional approval could benefit both family practice and rural health practitioners and may likely influence the way in which private insurance providers respond. The legislation would replace the current Medicare practice of paying doctors on customary, prevailing and reasonable fees. It would be based, instead, on a standard which includes the total costs of services provided, including education, equipment, and value of technical skills of the physician. Health experts believe that this would lead to greater emphasis on office consultations and preventative measures, and less on diagnostic and invasive procedures.

Another major social issue reflecting our lack of planning for the future is housing for the elderly. It has been a long established fact that older individuals in the U.S. value their freedom, privacy, and the opportunity to live independently from family and friends. For the elderly, medical care and housing arrangements are intertwined, thus, long-term care for older adults, many with chronic illnesses, will demand attention near the beginning of the next century.

Finally, the rural elderly have limited or no access to public transportation through they often live some distances from medical and shopping services. Lack of public transportation results in more rural elderly owning or having access to a car. When they are required to give up driving a car because of problems associated with advancing age, their freedom and independence is inhibited. It is this inhibition that leads to an increasing dependence on caregivers, who are most often female relatives or friends. To avoid nursing home placement, the average female caregiver will spend approximately the same number of years (18) caring for an aging parent as the years (17) needed for child rearing (Rieken, 1989).

AGRICULTURE AND NATURAL RESOURCES RESPONSE TO RURAL CONCERNS

Although the agricultural and natural resources systems can proclaim that issues involving rural youth, families, the elderly, and communities "are in someone else's backyard, so why worry," it should not be overlooked that the development of rural people plays a vital role in shaping the future of our nation's human resources. Those same human resources are students in colleges of agriculture and natural resources. They become researchers and scientists in our universities and agribusinesses, and producers of foodstuffs for international trade. The quality of life on farms and in rural communities determines whether families can retain the fierce independence and pride, long a tradition in rural areas.

Families are sometimes blamed for the moral corruption of our youth. Kunisawa (1982) wrote:

> The inability of the family and the community to teach its children traditional values is a leading contributor to the social decay we face at the end of the twentieth century.

When a unit of our society is so blamed, yet seen as the way for solving problems of the world, it is unconscionable for the family unit to be ignored in our academic, research, funding, and policy arenas.

BIBLIOGRAPHY

Boodman, S. C. 1989. "Newark Hospital Battles the Twin Plagues of Poverty." *Minneapolis Star Tribune* September 26:2.

Cowan, A. L. 1989. "Poll Finds Women's Gains Have Taken Personal Toll." *New York Times* August 21:1.

Edelman, M. W. 1987. *Families In Peril.* Harvard University Press.

Estrada, L. F. 1988. "Anticipating the Demographic Future." *Change* 20:14–19.

Gillis, W. R., and Schaffer, R. E. 1987. "Matching New Jobs to Rural Workers." *Rural Development Perspectives* October:19–23.

Glasgow, N., and Beale, C. L. 1985. "Rural Elderly in Demographic Perspective." *Rural Development Perspectives* 2:22–26.

Glazer, S. 1989. "Dropouts: An F For Education?" *Congressional Quarterly's Editorial Research Reports* April 21.

Haas, Cliff. 1989. "Congress Might Drop Catastrophic Illness Insurance For Elderly." *Minneapolis Star Tribune* October 1.

Hennon, C. B., and Brubaker, T. H. 1988. "Rural Families: Characteristics and Conceptualization, Families in Rural America: Stress, Adaptation and Revitalization." *NCFR.*

Hennon, C. B., and Marotz-Baden, R., eds. 1987. "Rural Families: Stability and Change." *Family Relations* October.

Johnston, L. D., O'Malley, P. M., and Bachman, J. G. 1985. "Use of Licit and Illicit Drugs By America's High School Students." National Institute of Drug Abuse.

Koff, T. H. 1988. *New Approaches to Health Care For An Aging Population.* Jossey-Bass Publishers.

Kolata, G. 1989. "AIDS Is Spreading in Teen-Agers, A New Trend Alarming to Experts." *New York Times* October 8.

Kramon, G. 1989. "Small Business Is Overwhelmed by Health Costs." *New York Times* October 1:1.

Krout, J. A. 1986. *The Aged in Rural America.* Greenwood Press.

Kunisawa, B. N. 1988. "A Nation in Crisis: The Dropout Dilemma." *NEA Today* January:61–65.

Levine, A. E. 1988. "Toward a National Service Program." *Change* September/October.

McGranahan, D. 1986. "Crime and the Countryside." *Rural Development Perspectives* February:2–8.

Minneapolis Star Tribune 1989. " 'Latchkey Kids' Drug Use Called Higher" September.

———. 1989. "EPA: Most Exposure to Toxic Air Is Indoors" September 29:15

———. 1989. "U.S. Kids Getting Fatter, Less Fit, AAU Finds in Decade-Long Study" October.

———. 1989. "Thousands Protest Housing Shortage" October 8:14.

———. 1989. "Life-expectancy Gap Widens Between Blacks and Whites" October 9.

———. 1989. "Survey Finds Third of Homeless Dependent on Drugs or Alcohol" October 11:7.

———. 1989. "U.S. Hispanic Population Has Risen 39% Since 1980" October 14.

Morrissey, E. S. 1986. "What Makes Poverty so Intractable in High-Poverty Nonmetro Counties?" *Rural Development Perspectives* February: 24–28.

Newsweek 1989. "Growing Up Against the Odds" September 11.

———. 1989. "The Elderly Duke It Out" September 11.

Neiger, B. L., and Hopkins, R. W. 1988. "Adolescent Suicide: Character Traits of High-Risk Teenagers." *Adolescence* Summer: 469–475.

Newcomb, M. D., and Bentler, P. M. 1988. *Consequences of Adolescent Drug Use.*

New York Times 1989. "Home Ownership Found To Decline" October 8:16.

Rathbone-McCuan, E. and Havens, B. eds. 1988. *North American Elders: United States and Canadian Perspectives.* Greenwood Press.

Rieken, J. L. 1989. "Aging and the American Family." *Journal of Home Economics* 81:53.

Roche, M. P. 1985. *Rural Police and Rural Youth.* University Press of Virginia.

Ross, P. J., and Morrissey, E. S. 1987. "Two Types of Rural Poor Need Different Kinds of Help." *Rural Development Perspectives* October:7–18.

Russell, C. 1989. "Health Issues Mattered in 1988 Votes for President." *Washington Post Health* October: 3:5.

Safire, W. 1989. "Goodbye, Bedroom Communities: Rural Is New American Way." *New York Times* October.

Select Committee on Children, Youth and Families, House of Representatives 1987. "Children and Families In The Midwest: Employment, Family Services and the Rural Economy." U.S. Government Printing Office.

Skrzycki, C. 1989. "Tough Times Confronting Older Workers." *Washington Post Business* October 2: 1,16.

Straub, L, and Walzer, N. 1988. *Financing Rural Health Care.* Praeger.

Swoboda, F. 1989. "45% of Black Children in U.S. Poor, Study Finds." *Washington Post* October 2:4.

Tolchin, M. 1989. "Medicare Change Is Seen Altering Fees For Doctors." *New York Times* October 13.

W. T. Grant Foundation Commission on Work, Family and Citizenship. 1988. "The Forgotten Half: Pathways to Success for America's Youth and Young Families." Final Report, November.

W. T. Grant Foundation Commission on Work, Family, and Citizenship. 1988. "The Forgotten Half: Non-College Youth in America." Interim Report, January.

Yogman, M., and Brazelton, T. B. 1986. *In Support of Families.* Harvard University Press.

10

The Biosphere, Future Generations, and the University

RUSSELL W. PETERSON

The biosphere, the habitat of all known life, is increasingly threatened by the cumulative activities of ever more humans. This mounting problem risks the quality of all life, but particularly that of future generations. Universities especially have a moral obligation to look after the rights of such generations. To fulfill this obligation, universities need to define the current global interdisciplinary predicament and its resolution and to educate and motivate tomorrow's leaders to be able and willing to apply this knowledge to safeguard posterity. Such long-range programming will also pay off to today's university students over the five to six remaining decades they will be extant. And the players who exercise the vision and boldness to pursue this course will gain the satisfaction of participating in a great cause.

Clearly we have a moral obligation to do so.

It took several billion years and an infinite number of biochemical experiments producing along the way amino acids and double helix chains before Earth, with the aid of solar energy, was able to put together the current biosphere. This Earth-encircling assembly of plants, animals, waters and soils, recycling and regenerating, evolving its vital components, and creating and regulating its life-support systems is a magnificent creation. But today it is under serious attack by a powerful ecological force that is altering its self-regulating mechanisms, intervening in its evolutionary process and precipitating its sixth great extinction crisis. We humans are that force, over five billion of us, assaulting the biosphere with ever more machines, chemicals, bombs and waste. What kind of stewards of planet Earth are we!

Last year I chaired an international symposium marking the 100th anniversary of the National Geographic Society. Each of the 21 internationally renowned speakers discussed what had happened in his or her field during the past 100 years and what might happen in the next few decades. Their message was clear. We are facing an unprecedented series of crises which call for urgent action if we are to avoid severe economic decline and social disintegration.

Our preoccupation with the present blinds us to that fact. The impact of our actions must be weighed over time. Consider the following:

- The world population is growing faster in absolute numbers than ever before, now adding its sixth billion over only eleven years. At its current rate of growth, it will triple during the life of today's university students. The United Nations' Fund for Population Activities that had previously projected that the world population could be leveled off at ten billion announced in May 1989 that because of inadequate family planning and socio-economic development programs, the population is unlikely to be leveled off before fourteen billion.
- The world economy is growing even faster—more than five-fold since World War II—and projected to grow five to ten-fold over the next fifty years. This growth will bring a commensurate increase in humans' impact on the biosphere, especially by affluent countries such as the United States where one person already uses 40–50 times as much of the world's resources as a person in the poorest countries.
- Approximately one billion people now live in absolute poverty, a condition well below any definition of human decency. Among them, 40,000 children die daily, and many more are stunted from malnutrition, polluted water and childhood diseases regularly avoided in more affluent societies. No other current human tragedy approaches this in severity.
- The United Nations Children's Fund (UNICEF) in its 1988 "State of the World's Children" reported, "For approximately one-sixth of mankind, the march of human progress has now become a retreat and, in many nations, development is being thrown into reverse. After decades of steady economic advance, large areas of the world are sliding backward into poverty."
- The World Food Council of the United Nations in its 1988 "Global State of Hunger and Malnutrition" reported, "Earlier progress in fighting hunger, malnutrition and poverty has come to a halt or is being reversed in many parts of the world."
- Lester Brown, in *World Watch* magazine, stated that in 1988 United States grain production fell below consumption for the first time in modern history, and that in July 1989 the U.S. Department of

Agriculture estimated that the 1989 harvest would be 13 million tons below consumption of 1,164 million tons. He reported that world grain stocks have fallen to one of the lowest levels in decades—a little more than to fill the pipeline from field to table. As a result, world grain prices between 1986 and 1988 increased almost 50%. He predicts that another low U.S. harvest, like in 1988, before the stocks are rebuilt could lead to a doubling or tripling of prices, bringing on catastrophe.

- Although world grain production more than kept up with population growth during the 50's, 60's and 70's—the production per person increasing nine percent per decade—in the 80's it fell by two percent and is expected to fall by seven percent in the 90's. (*World Watch*, 1988. September-October).
- Earlier increases in grain production came in part from bringing more land under cultivation. Now, with less land potentially available and suitable for grain production, and with accelerating conversion of existing cropland for urban and other growth, further increases in production will have to come from higher productivity per acre.
- While the world cries out for more food, fiber and wood, excessive cropping, grazing and deforestation already are causing the most rapid desertification, soil erosion, wetlands destruction, groundwater depletion, and species extinction in history.
- Hanging over all life is the threat of nuclear weapons which, some afternoon or evening, by accident, mistake, equipment failure or deliberate act, could not only evaporate, blast and irradiate to death much of the earth's life, but could decrease sunlight, temperature, rainfall, and the stratospheric ozone layer, devastating agriculture even in noncombatant countries and bringing global nuclear famine. The high probability of such result from a nuclear exchange, first reported in 1982, was confirmed in 1988 after six years of study by the Scientific Committee on Protection of the Environment of the International Council of Scientific Unions. A scientific consensus on this view is now emerging. The United Nations General Assembly on December 7, 1988, with 145 ayes, zero nays and nine abstentions including the U.S. commended a report expressing the above view by its Group of Consultant Experts to the attention of all Member States. Of especial concern is that the leaders of most of the NATO nations have almost completely ignored this super threat to the biosphere, presumably because acknowledgment of it would make any policy to use nuclear weapons under any condition clearly immoral and suicidal.
- The release of the so-called greenhouse gases into the atmosphere, especially carbon dioxide from the burning of fossil fuels, threatens

a global warming that could markedly affect agriculture and raise ocean levels, thereby flooding coastal areas. The record high temperatures in the 1980's with accompanying drought triggered worldwide concern. Although there is considerable debate over the implications of the 1980's heat wave, many scientists believe it is probable that the warming resulted in part from release of greenhouse gases over past decades. In any event the warming provided a preview of what is anticipated for the decades ahead by many scientists around the world.

- There is little doubt today that the release of chlorofluorocarbons from cooling machines and plastic foams depletes the life-protecting ozone layer in the stratosphere. Even the manufacturers who produce the chlorofluorocarbons accept this and are proceeding in agreement with their governments to eliminate the production of the chlorofluorocarbons over the next decade. Nevertheless, those already released will persist in the stratosphere depleting its ozone for decades.

- Our way of life, including agriculture, has been fueled over the years by a growing supply of energy, especially oil. Now it appears that the world production of oil will peak in 10–15 years and then gradually decline to zero. This indicates that over the lifetime of today's university students, the world will use 80% of the oil it will ever produce. And by humanity's own choice, the world production of nuclear energy is also likely to peak in 15–20 years and then gradually decline. As a result in the downturn in production of these two energy sources, energy prices will probably rise markedly, impacting especially hard on energy-intensive agriculture.

- Over many years, nations have dumped their wastes into their own and their neighbors' air, water, and land, creating poisonous dumps, polluted streams, acid rain, urban smog, unsafe drinking water, contaminated beaches, and nuclear wastes, threatening all life and costing today's and tomorrow's taxpayers hundreds of billions of dollars to pay for this past negligence. And the practice continues.

- Escalating debt at consumer, corporate and national levels here in the United States and in many other countries has dried up the investment capital required to cope with the cumulating predicament. This coupled with political commitments to no taxes in order to get elected leaves only token amounts available for saving the world.

- An especially outrageous display of lack of concern for the future is the present action of our federal government in reducing the annual deficit by spending on current expenses the tens of billions of surplus dollars from social security taxes that were supposed to be saved for payment of pensions when the "baby-boom generation" reaches retirement age.

- Exacerbating the shortage of U.S. resources necessary to deal with the critical global problems is the maintenance—even though we are at peace—of a military machine thousands of times more powerful and destructive than any that ever existed before—even in wartime. Today 52% of our federal income tax goes for military and military-related costs. Only two percent is spent on education and two percent on natural resources and the environment. One-third of all U.S. scientists and engineers are engaged in military-related work. Twenty-six million men, women, and children depend on the Pentagon for their livelihood. A cozy quandrangle among the Pentagon, committees of the Congress, industry, and scientists perpetuates this excessive use of resources.

The cumulative impact on the biosphere that I have described provides an awesome challenge for today's and tomorrow's leaders and a super menace to the quality of life of future generations. This growing global threat has been recognized for years by people of vision, the so-called doomsayers.

Now it appears that the general public and many leaders are finally awakening to this serious predicament. President Bush, President Gorbachev, former Prime Minister Thatcher, and other world leaders have spoken of the need to protect the environment but so far have provided no additional resources.

When I look back at how far the world has come over the 55 years since I enrolled as a freshman at the University of Wisconsin, I see that amazing stream of remarkable accomplishments by Homo sapiens. It is hard to believe that we can't cope with today's challenges.

All is not gloom and doom. What is needed are more people, and especially leaders, who think comprehensively and globally, who weigh the impacts of today's decisions on the quality of life of future generations, who are able to make responsible choices among alternative futures, and who gain job satisfaction primarily from contributing to the welfare of others. It is time, if I may paraphrase President Kennedy, to "Ask not what your biosphere can do for you, but what you can do for your biosphere."

Toward this end, educators must emphasize more extensive and effective teaching about the whole. In light of our current understanding of the implications to all walks of life of the global interdependence of things, no one whose training is solely in a traditional discipline can justifiably claim to be adequately educated.

One early champion of seeing things whole was the University of Wisconsin's renowned biologist, Aldo Leopold, born 102 years ago. He put it this way: "All the sciences and arts are taught as though they were separate. They are separate only in the classroom. Step out on the campus and they are immediately fused."

My lifetime of careers in many fields has convinced me that it is more the rule than the exception that people enter the most influential positions in our society ill prepared for the breadth of their assignments. What is required are securely funded discrete Colleges of Integrated Studies, as I recommended in an article in the Spring 1988 issue of the University of Wisconsin *L&S Magazine*. Such institutions, through rigorous undergraduate and graduate training, could produce "professional generalists" who thought comprehensively, globally and long term; who were enlightened by the historical record, the arts and humanities; who understood the interdependence of all things; who integrated the technical, economic, social, environmental and political variables involved; and who assessed the long-term impacts and feedbacks of their decisions. This is the training required for the most important jobs in our society. Yet no institution provides it.

There is great interest around the country, yes around the world, in such a proposal. This interest, however, is held by a minority of the faculty on any one campus, albeit an enlightened and highly dedicated minority. I personally know dozens of such faculty around the world. They tell me there are thousands like them. They are struggling to make the college curricula pertinent to the real world. They understand the mounting global crisis we Homo sapiens have created as our global society has blindly moved into the future. They know how we can face up to this global predicament. They are a great asset. We must provide them the encouragement and means to bring their enlightenment to bear more effectively on teaching and sparking more students and in providing the ideas needed for humanity's mutually assured survival.

A separate College of Integrated Studies would provide them the environment and freedom in which they could flourish. Resistance to the idea of an integrated studies discipline arises among members of the traditional disciplines who see the proposed meta-disciplinary program impinging on their territory, competing for funds and clashing with the ideology that advanced teaching is best served by practicing it in ever narrower fields. Certainly the action of universities, as knowledge exploded, to develop narrower and narrower fields of specialization has been highly successful in advancing the frontiers of knowledge. It will continue to prosper. And the more recent proliferation of interdisciplinary programs that pool the knowledge of the brains of specialists from several disciplines have furthered the understanding of the global interconnection of things and helped to find solutions to some critical problems.

But I believe there is a great need for the professional generalist not only in research where his lone brain can operate effectively at the interface of several disciplines and benefit from symbiosis among them, but also in leadership positions where he can integrate the increasingly narrow slices of knowledge presented and make intelligent choices among alternative futures. Probably in agriculture, life sciences and natural resources there is fertile soil for this new Renaissance institution—for this pioneering beachhead in academe.

Consider the many issues I discussed earlier that threaten the biosphere and future generations. They deal with plant, animal and human life, agriculture, the air, water, land, sun and energy and call for expertise in both the physical and social sciences and enlightenment by the humanities.

Why not convert the University of Wisconsin's two-year Integrated Liberal Studies program into a four- or five-year rigorous undergraduate program for Renaissance students and couple it with a graduate school for professional generalists dedicated to coping with the global problems I have described? Such an institution could provide the creativity, training, vision and boldness necessary to develop solutions to the global crisis, to define what is needed to carry them out and through an extension service teach and motivate society to make it happen. Maybe the Governor and some leaders in the Legislature could help make it happen.

So far there is no evidence that the top leadership of the University of Wisconsin appreciates the great need for professionals with breadth of knowledge. Their decreased support for inter-disciplinary programs in recent years attests to the problem.

Dr. Charles Benbrook, Executive Director, Board of Agriculture, National Research Council, reported that the federal government was making $250 million available specifically for multi-disciplinary research. Why not set-up the institution I described and go after five to ten percent of that money?

Let's consider from today's perspective examples of what needs to be done to cope with the global predicament. First, it's essential that population growth be stopped as soon as possible. Even at the current population level (5.3 billion) the rate of degradation of the resource base on which life is dependent is so great that it is questionable whether humankind could provide a decent quality of life for this number on a sustainable basis. All developed nations have lowered the births per mother to less than the replacement level of 2.1. But in Africa, Latin America and South Asia birth rates markedly exceed death rates. This condition cannot last for long. If the birth rates are not lowered, the death rates will eventually rise to limit population growth. In May 1989, as I stated earlier, the United Fund for Population Activities raised to 14 billion its projection of where world population might be leveled off, based on their current estimate of the effort the world community is likely to put into reducing birth rates.

The current more pessimistic view stems in part from the abandonment during the Reagan Administration of the United States' bipartisan world leadership in furthering family-planning assistance in the developing countries. Among other things, the U.S. cut off all of its major funding of the two most effective international family-planning agencies. The Bush Administration thus far has not changed this policy. By withholding these funds, the U.S. has denied millions of poor women the means of avoiding unwanted pregnancies thereby increasing abortions and the birth into absolute poverty of large numbers of children who will suffer and die from hunger and neglect.

A prime choice for U.S. citizens is to put our democracy to work to turn our government around on this issue, convincing them to markedly increase family planning and other socio-economic aid and to fund research on new contraceptives. By reducing its population growth, a nation will have less difficulty in providing for the needs of its people and in preserving the natural resource base on which their livelihood is dependent. This is illustrated by the Population Crisis Committee's human suffering index which shows an almost perfect correlation (a Pearson coefficient of 0.83) with a nation's rate of population growth.

Nearly all people everywhere aspire to using more resources. This is readily understandable for those who are struggling to reach subsistence levels and especially for the hundreds of millions in Africa, Latin America, and South Asia whose livelihood has been decreasing over the past decade. It is also readily understandable for the poor who live in the affluent societies such as the U.S., where the poor are getting poorer while the rich are getting richer. But it is also true that the affluent nations are counting on a continuing growth in their prosperity. The U.S., Japan, Western Europe, and other developed nations are planning and expecting a continuing growth in their already huge production of goods and services (GNP) of at least three percent per year. That would mean a doubling in only 23 years, or about a five-fold increase during the remaining lifetime of today's university students.

At the same time, the USSR and other centrally planned economies are moving toward freer market economies that promise a more rapid growth. China with its 1.1 billion people doubled its GNP over the past nine years. Such economic growth will have a tremendous impact on the environment and on the rate of depletion of natural resources. How long it can be maintained is highly problematical. Exponential growth cannot go on forever even in agriculture. Sooner or later it will run up against limits.

Growth in the production of oil illustrates the point. From 1870 to 1930 the production of oil in the U.S.'s contiguous 48 states increased exponentially, doubling every eight years. Then, as King Hubbert projected in the 1930's, the rate of increase fell off until production peaked and started downhill in 1970. It will continue to decline to zero. The U.S. is temporarily solving this problem by importing oil. But the world production of oil will peak in about fifteen years and start its decline toward zero no matter what the forecasts for future production call for.

The fact that the annual growth in GNP among the developed countries has fallen over the past three decades may indicate that the world is running up against limits to growth. Although Chambers of Commerce and many economists consider such statements heresy, it is time for more concerted effort to face up to limits to growth. To reduce growth in use of resources, our society needs to concentrate on extending the life of products we use, recovering and recycling materials, renovating rather than replacing buildings, furthering the use of bicycles and mass transit and developing industrial processes and life styles that avoid pollution and waste.

It is hard to believe that we can ever manage our economy and environment to provide for a sustainable, decent livelihood globally until we develop a more adequate tool than GNP to measure economic progress. GNP weighs the consumption of natural resources as progress, not recognizing that such consumption entails the depletion of natural capital. Under the banner of economic development we clear the forests, erode the soils, fill in the wetlands, pollute the air and water, hunt and fish to extinction, deplete the oil wells and exhaust the aquifers to enhance the GNP, but in the process assure humanity's insecurity.

To further the world's security, it is essential that we face up to the future development and use of energy, so vital to our way of life, but a principal cause of degradation of the environment. Of especial concern today is the burning of fossil fuels. It causes urban smog and acid rain and other air pollutants that are devastating forests, killing fish, and corroding buildings; it leads to oil spills such as the recent catastrophic event in Alaska; and above all, it produces the greenhouse gas, carbon dioxide, which threatens serious alteration of global climate.

By far the best route to cope with these problems is to further the soft energy path, that is, to develop more efficient use of energy and renewable sources of energy. It is also a key to reducing energy costs, making industry more competitive, reducing oil imports and reducing the need for nuclear power with its attendant nuclear waste and nuclear weapon proliferation problems.

Amory Lovins, a principal proponent of this so-called "soft-energy path," illustrates the potential for energy efficiency by showing, for example, how through the installation of a type of lighting now being marketed, 25% of all the electricity used in the United States could be saved. This would permit the avoidance of building 120 one-thousand-megawatt power plants at a construction cost of several hundred billion dollars, as well as the avoidance of the annual cost to operate such plants of about 30 billion dollars. For example, one can reduce acid rain, urban smog and other pollution, prevent the production of one ton of carbon dioxide and obtain the same level of lighting by replacing one 75-watt incandescent lightbulb with an 18-watt fluorescent bulb now available.

Major savings and environmental protection can also be achieved by furthering the marketing of more fuel-efficient automobiles. Under pressure from General Motors and Ford, an earlier promising effort toward this objective was stymied by our federal government. The Office of Technology Assessment now estimates that under current regulations the use of fuel by cars and other light vehicles in the United States will be slightly higher in the year 2000 than now. Modest improvements in miles per gallon will be overtaken by an increase in cars on the road.

Technology exists to increase markedly the current requirement that manufacturers' new car fleets average at least 27.5 miles per gallon. Some commercial models now give over forty miles per gallon and some experimental models exceed 100 miles per gallon. Consumers' increasing demand for the heavier gas guzzlers which are more profitable for auto manufacturers counteracts efforts toward higher efficiency. Worldwide, more attention must be given to alternatives to the automobile. A September 1989 *Worldwatch Paper* (#90), "The Bicycle: Vehicle for a Small Planet," shows one way.

A major broad-based soft-energy program launched by our government in the 1970's was scuttled in the 1980's. In spite of all the concern and all the speeches about global warming, air pollution and oil imports, nothing has yet been done by our federal government to revive this most promising route toward mitigating these problems. The sun is the natural resource which promises to provide the key to our energy future. Already wood and other forms of biomass, products of photosynthesis, provide more energy than nuclear plants at prices competitive with oil and gas. The opportunities for further development are substantial.

Of especial interest are the renewable energy technologies that capture solar energy more directly, such as solar thermal and photovoltaic technologies. Solar thermal uses mirrors to focus sunlight on a boiler to convert water to steam which drives a turbine to produce electricity.

Luz International has developed a solar thermal process which uses natural gas to heat the water when the sun isn't shining, thereby providing electricity around the clock. They have already produced electricity at 11.4 cents per kilowatt hour as compared with 12.4 cents for the newest nuclear plants. The first of five 80-megawatt plants ordered by Southern California Edison Company will be started up in the Mojave Desert in December 1989. It should produce electricity at 7.9 cents per kilowatt hour. They estimate that a 160-megawatt plant could produce at 6.3 cents/KWH, the same as plants based on clean coal technologies. It is estimated that solar thermal plants could fulfill one-quarter of the electricity needs of the United States.

An even more promising development is the photovoltaic device which converts sunlight directly to electricity by shining it on certain materials such as silicon. Photovoltaics are already used to power watches, small computers, electricity generators in remote locations and satellites in space. Major research on this development is under way in Japan, Europe and the United States. Advances have been coming rapidly. It is projected that this technology will be cost-competitive with nuclear energy by the year 2000.

A promising longer-range development is to use the direct current electricity produced by the photovoltaic generator to electrolyze water, producing the clean-burning fuel, hydrogen. This development merits major research effort. It harvests the renewable energy from the sun, can be located in the most sunny areas such as deserts, can operate intermittently with sunlight, provides a portable fuel for transportation and does not contribute to global warming, urban smog, acid rain or hazardous wastes.

A sign of the times is the current installation of a plant for manufacturing photovoltaic cells on the site in Wackersdorf, West Germany, where a plutonium reprocessing plant was abandoned after spending $1.3 billion.

The largest current user of solar energy, agriculture, will need to receive major attention to improve its efficiency while reducing its serious impact on the environment. The massive application of chemical poisons and fertilizers, which has increased production, has harmed other parts of the biosphere and has consumed huge quantities of rapidly depleting oil.

The National Research Council of the National Academy of Sciences issued a report on September 7, 1989, that supports a growing movement of farmers and consumers who advocate a fundamental change in agriculture away from its heavy reliance on chemicals. Dr. Charles E. Hess, Assistant Secretary of Agriculture, said he would seek new funds to support alternative methods of agriculture which range from pure organic farming to new techniques of low-chemical pest management, crop rotation, tillage that reduces erosion and genetic improvement to plants to resist pests and disease.

California is out front in substituting alternatives to pesticides. Since its hot, dry climate is well suited to the new techniques and since it produces about half of America's vegetables, fruits and nuts—the products especially vulnerable to pests—it is likely that California will be the pacesetter in alternative pest management. Supermarkets, in response to consumer demand, are using their large purchasing power to further less chemical-intensive farming. It may well become the dominant form of agriculture by the year 2000.

Another assignment for agriculture is to reforest the earth. This is important not only to increase absorption of the greenhouse gas carbon dioxide, but also to reduce soil erosion and flooding, replenish ground-water supplies, save endangered species and provide biomass for fuel.

The current emphasis on the development of genetically improved plants and animals has the potential for great good but also substantial harm. Caution is called for. These new products, like new chemicals, should be presumed guilty until proven innocent. Society needs to be more concerned about the long-range impact of the choices made by the world's movers and shakers in selecting technology to satisfy its needs. We should not leave the choice of technology strictly to consideration of the financial return to be gained by those who invest in it. Adam Smith's "invisible hand" might beckon us forward today, but an "invisible foot" could kick us in the future as it did with nuclear energy and pesticides.

Water has played a principal role in the huge growth in agriculture, industry and municipalities over past decades. Limits to its additional use are now appearing all over the world. Wasteful and inefficient use, especially in irrigation—encouraged in many places by large government subsidies—is a prime cause of the problem. Drying lakes, shrinking rivers and depleted aquifers attest to the problem. It calls for political action to make water users bear the full costs of their water and for the development and application of technology to further more efficient and less wasteful uses.

As I mentioned earlier, a major threat to the biosphere is the creation, production and deployment by the U.S., USSR, Britain, France, and China of nuclear weapons, the common super-enemy of all life. The nuclear ordinance on one U.S. Trident II submarine, for example, could ignite enough cities to produce sufficient smoke to block out sunlight, lowering temperatures and devastating agriculture and plant and animal life over wide areas of the world. Scientists have now thoroughly confirmed the potential for this nuclear global cooling and its catastrophic biological consequences. Since political leaders ignore this finding, it seems essential that those in agriculture and life sciences should ring the alarm.

The tens of thousands of nuclear warheads deployed around the world and the hundreds of thousands of troops trained to use them make a catastrophe due to equipment failure, human error or unauthorized firing disturbingly possible. The accidents caused by humans at Three Mile Island and Chernobyl, the shooting down by mistake of a commercial airliner by a modern, sophisticated warship, an oil tanker straying out of a ten-mile-wide channel in Alaskan waters to hit a well-known reef, and the alleged suicidal blowing up of a 47-man gun crew on the battleship U.S.S. Iowa by the gun crew captain illustrate the potential for humans to frustrate the most carefully designed safety programs.

Until the distant day when nuclear weapons might be banned globally, activists should work to lower the number of nuclear weapons to some small fraction—say five percent—of today's inventory and provide multiple safeguards against inadvertent use of the remainder.

The USSR and the United States built their nuclear arsenals under the perception that the other was their mortal enemy. This two-sided cold war climate has now markedly cooled as President Gorbachev has demonstrated his determination to cool it and Presidents Reagan and Bush have agreed to reciprocate. Evidence of the change is the statement by the recently retired chairman of the Joint Chiefs of Staff, Admiral William J. Crowe, that he does not consider the Soviets as his enemy, but as a potential adversary. And in June, the Pentagon issued the "1989 Joint Military Net Assessment" of the Joint Chiefs, stating that the Soviets' primary concern is the security and integrity of their homeland. To deter war with the West is their primary security objective which they have codified in their military doctrine. The Soviets' build-up of nuclear missiles, the document states, was done "as a more comprehensive form of deterrence rather than as an expression of the belief that they could win a nuclear war." With this new climate, we should be moving more rapidly to reduce the number of nuclear weapons, the super-threat to the biosphere.

All of the issues I have discussed are interconnected, impact on the biosphere, and are vital to the quality of life of future generations. They call for more understanding, more research, more comprehensive policies, more long-range planning and much more commitment of financial resources.

No token appropriation will suffice. As a starter, the U.S. needs to invest at least $50 billion additionally per year on global programs to further family planning, develop better contraceptives, push more efficient use of energy and renewable sources of energy, reforest the earth, promote alternative forms of agriculture, develop industrial processes that avoid the production of hazardous wastes, recycle and recover materials, and clean up the inherited pollution. Other developed nations also need to make a major commitment. The funds for this mutually assured survival program could safely come from the military budgets now dedicated to our mutually assured destruction.

Lester Brown, in *State of the World 1988,* sketched a plan for the world to take 16% of the $900 billion it is spending per year on its military functions and allocate it to ecological security programs. Such a comprehensive worldwide strategy and program are essential. No single nation or small number of nations can do the job alone. The United Nations Environment Program has been showing the way by organizing international conferences to deal separately with phasing out of chlorofluorocarbons globally, controlling international disposal of waste and global warming. But what is required is a grand strategy and a global commitment to deal simultaneously with the major ecological, economic, military, social and political forces at work. This is a job for the United Nations Security Council to initiate. For it to act, however, will take sustained and bold joint leadership by the U.S. and the USSR.

What is required is foreign to our current way of operating—which involves piecemeal policies and programs focussed on the near term, made urgent by promotion through photo opportunities and the nationwide news media and charged in large measure to future taxpayers. The growing concern of people around the world about the threats to plant earth as shown by polls, the increased frequency of international environmental conferences and protocols and the recent chorus of pro-environment statements by heads of state are encouraging. But so far they have led to little action.

It would be helpful if we would stop supporting politicians who insult us as citizens of affluent America by appealing to penny-pinching instincts and by promoting "borrow today—pay tomorrow" policies while the world desperately needs our leadership in investing in a healthy sustainable biosphere. The trends are too threatening, the stakes too high and the rate of change too rapid to tolerate this approach much longer.

The United States should be leading the way out of this global predicament. It isn't. You and I and thousands like us at the local level all over America need to see that it does—need to put our knowledge of political science to work.

Students, faculty and administrators should pay more attention to their role as citizens of the world. Although it is important to be involved with local issues, the global interdependence of all components of the biosphere makes a world view a prerequisite to effective citizen action locally and globally. Universities are too quiescent on the big global issues, especially the students. Students should be crying out, demanding action. They are the recipients of a sad legacy and will be among the leaders for the next half century, for better or worse. They can become part of the solution or add to the problem. What choice they make will depend in large measure on their university experience.

Time is running out. We must stop the on-going desecration of the biosphere and work toward the resurrection of this only known habitat of life. The world has the necessary resources and the knowledge of what needs to be done. What is missing is the will and the commitment to do the job. Somehow we must regulate the conflicts of ideology, religion, socio-economic status and politics to the sidelines while we get on with the big game—the survival of life on earth.

It would help if we put being a world citizen ahead of being a Republican or a Democrat. It would help if each of us could say as Mahatma Gandhi did, "I am a Hindu, a Muslim, a Christian, a Jew and so are you." It would help if each of us would weigh the tragedy of 40,000 children dying each day from avoidable causes and resolve to do unto others as we would have them do unto us. It would help if more students began to recognize that catastrophic change can occur in their lifetimes unless they get involved in changing the trends of today.

Each of us can make a difference. Universities can make a huge difference. Universities can spark a renaissance of learning about the biosphere and a commitment toward goals beyond oneself. What greater moral mission is there than working toward the welfare of life after us. The University of Wisconsin's agenda for the next 30 years could show the way. As Goethe wrote:

Whatever you believe you can, or dream you can, try it.
Boldness has genius, power and magic in it.

III

Program Futures for Agriculture and Natural Resources

11

Charting a Course for 2020

WAVA G. HANEY
DONALD R. FIELD

20–20 Vision? As the Steering Committee for Strategic Planning began to plan for the future of agriculture and natural resource education in the University of Wisconsin System, they reflected upon the arguments made by the authors in this volume and experimented with views of how these fields will be functioning 30 years in the future, anticipating that the world will be a very different place in the year 2020. Many of the trends which have begun to emerge will continue to be evident, but other changes may be more difficult to project.

Demands on the use of the natural resource base will almost surely increase in ways we cannot even imagine today, and principles of stewardship of resources will permeate private and public decision making. Public opinion will also continue to affect the way that producers and processors operate.

In a finite, interdependent, and rapidly changing world, the maintenance of strong food, fiber, and forest production systems and the protection and wise use of our natural resources have become an imperative. However, agriculture and natural resource programs find themselves in a time of uncertainty. For example, where once American agricultural and forest production was unchallenged in a U.S./North American market place, now they have become integrated into a world market. Thus, as economic policy in the European Community and eastern Europe unfolds, agriculture and forestry markets for those parts of the global market will change and new product demands will surface.

Similarly the interdependency of agriculture, forestry, and fisheries production practices with conservation, preservation and recreational use of land and water systems has now linked farmers with environmentalists. Likewise, food safety concerns and groundwater contamination are changing contemporary thinking about food and fiber production with the balance of nature. Genetic engineering, biotechnology, alternative agricultural practices, and sustainable rural development are simultaneously linked with global climate change, acid rain, groundwater contamination and extinction of species.

The farmer, forester, aquaculturist, energy developer, community specialist, warden, and park manager will be required to blend development actions in the context of natural resource conservation and enhancement.

Universities have an obligation to provide leadership for the changes which have been outlined. Inquiry into questions of production efficiency, profitability, technology, product safety, social viability and ecological sustainability, as well as the balance among these issues must be conducted. Through their research, teaching and extension activities, faculty and staff involved in agricultural and natural resources programs must be committed to framing and studying these challenging questions basic to the welfare of current and future generations.

Further, agriculture and natural resource students must not only be knowledgeable about a profession, but have developed critical thinking skills, a broad ecological, social, and economic perspective, and appreciation for the aesthetic. They must be able to write and speak effectively, to be sensitive to our cultural and natural heritage and our contemporary social institutions, and to converse in more than one language. Students of tomorrow must also be able to understand the interrelationship of their activity to an interdependent environmental world shared by food producers and consumers alike.

To address the issues and trends anticipated in this volume and to respond to the concerns expressed by the chapter authors, a set of goals for agricultural and natural resource programs have been identified. Each goal is buttressed with a justification statement synthesized from the previous chapters. The goals represent the collective recommendations of the Strategic Planning Committee for the future of agriculture and natural resource education in Wisconsin. The global statements were compiled by the editors.

GOAL 1: PROVIDE THE KNOWLEDGE BASE
TO ACCOMMODATE RAPID CHANGE

Strategies. Commitment to provide intellectual and scientific leadership: to improve the economic viability of the food, fiber, forest, and recreation systems; to protect, conserve, and enhance natural resources of the state, nation, and world; to sustain production efficiency in balance with stewardship of natural resources and maintenance of a safe food supply; and to support healthy families and livable communities.

Recommendation 1.1: Provide Leadership in Worldwide Research. *Provide domestic and international leadership in research on food and agriculture, natural resource and community viability issues.*

Research and education are basic to intellectual leadership. The historic *Wisconsin Idea* has been to do research in service to the people and extend it to the boundaries of the state and beyond. It is the teacher–scholar who expands minds, instills a quest for knowledge and creates the unique learning culture for the entire state. Teaching and extension are informed by research and in turn inform research.

If it is to serve society, the research process must span a spectrum of activity from basic to applied science. Research in the basic sciences is the long-term driving force of the system, but that base has been eroding. A renewed commitment to basic research is required to keep the System competitive in the biological and social sciences. Of equal concern, applied research and outreach mechanisms have not kept pace with state, national, or international needs.

The UW System must maintain excellent researchers by:

1. hiring and retaining the best scientific talent for high priority new and existing research areas, while phasing out areas of lower priority;
2. maintaining and rewarding scientific currency through career development opportunities for every active scholar in the System, with special attention given to faculty with limited or no formal research appointments; and
3. promoting and rewarding linkage among the campuses within the system among faculty in allied disciplines, utilizing especially the strong research base on the doctoral campuses.

The UW System must make adequate facilities and equipment available to its researchers in cost effective ways by:

1. seeking innovative ways to procure and maintain state-of-the-art laboratories, field sites and equipment by private- or public-sector partnerships, and
2. forming state, regional or national consortia in some fields.

The research system is constrained by adequacy of personnel, infrastructure and financial support. Maintaining support essential for scientific competence will become increasingly difficult and expensive. Meaningful participation in research requires that sufficient time be made available for that pursuit. Scientists must have access to the latest information by direct contact with colleagues within and outside the system, adequate library based information, and personal contacts through interpersonal exchanges such as scientific meetings. Additional support for significant research and professional interaction must be provided.

Laboratories, experimental field sites and operating farms, and the equipment needed in modern science are becoming increasingly difficult and expensive for the university system to buy and maintain. The unavailability of many expensive equipment items will increasingly limit the scope and quality of research and the accompanying training of graduate-level students. Deficiencies in facilities must be confronted and remedied. Innovative ways must be found to modernize laboratories and experimental field areas if university researchers are to bring contemporary quality scientific inquiry to their institutions. Difficult, complex decisions must be made by the Board of Regents about the desirability and affordability of attempting to achieve enhanced levels of research capabilities on all or an expanded number of campuses within the UW System.

Recommendation 1.2: Facilitate Basic, Applied, and Interdisciplinary Research. Maintain excellence along the basic-applied science spectrum.
To accomplish the commitment to intellectual and scientific leadership, the UW System research program must maintain excellence in critical areas to ensure the ultimate use of research findings to solve problems or create opportunities by:

1. maintaining the present strong base of individual, discipline-based research;
2. facilitating and developing multidisciplinary and problem-focused research, without loss of academic freedom, initiative, responsibility or accountability;
3. improving transfer of scientific, technological and policy findings to the class room, to the citizens of the state, and to the economic community; and
4. recognizing the increasingly inter-dependent nature of the world's agricultural and natural resources.

Research itself is an interactive system encompassing a spectrum of operational levels from basic sciences at one extreme, to adaptation and application at the other. There must be a constant flow of information along this continuum between basic and applied science, and the ultimate incorporation of findings into general use and practice in both public and private sectors. Knowledge within the scientific discipline must be advanced, and the public welfare and economic development of the state benefited.

One of the major challenges facing academic research over the next three decades will be the development of increasing flexibility in the conduct of research. The single faculty investigator research system based on individuals winning competitive grants has worked extraordinarily well on the UW–Madison campus; it must continue. But alternative modes of research and sources of funding must also be found.

Research needed to define and solve many of the problems in agriculture and natural resources requires expertise in a variety of social, biological, and physical science disciplines working collaboratively. Indeed, collaborative networks may span beyond the college and the campus. Mechanisms must be developed to recognize complex research problems and assemble interdisciplinary research teams to address them. It is reasonable for the state to support interdisciplinary approaches that deal with state problems of high priority. The faculty incentive and reward system must have sufficient flexibility to accommodate interdisciplinary research.

Problems impacting agriculture and natural resources are not confined within national borders. Environmental deterioration anywhere affects the whole planet. The forces that drive ecological degradation are complex, involving social, economic and political elements in addition to the obvious biologic and physical factors that are operating. Approaches to agricultural development and natural resource management that are not broadly interdisciplinary are doomed to failure. Developing countries will require research collaboration from our scientists to preserve their food and natural resource base, and to design alternative institutions and policies. Scientific assistance may well be the greatest contribution that we can make to the stability and improvement of life in these countries, and yet the proportion of university scientists with international experience is rapidly declining. Junior faculty are often actively discouraged from international involvement by their senior colleagues. We must accept the responsibility to help reverse these trends, through a greater effort to internationalize university research.

GOAL 2: PREPARE STUDENTS FOR LEADERSHIP IN TRADITIONAL AND EMERGING PROFESSIONS AND AS EDUCATED CITIZENS

Strategies. Commitment to strengthen curricula and learning environments consistent with new and evolving concepts of agriculture and natural resources, technologies and understanding of learning.

To sustain a leadership position, UW System colleges must offer and maintain curricula that reflect both rapidly changing technology and broader concepts of agriculture and natural resources, and a curricular planning process flexible enough to keep it current.

Maintaining a margin of excellence for its undergraduate and graduate students in agriculture and natural resources can be done by:

1. expanding curricula to incorporate new areas and more encompassing concepts;
2. delivering curricula in a manner that generates value questions and fosters a range of skills, yet enables students from varying levels of experience at entry to meet program goals;
3. providing opportunities for independent learning and leadership development;
4. recruiting and retaining well-trained faculty from diverse social and intellectual backgrounds to teach in new, evolving and existing areas (also see Goal 6);
5. developing faculty growth opportunities and reward structures that recognize good teaching; and
6. strengthening the development and utilization of new technologies and methodologies that enhance instructional quality, creativity, collaboration and efficiency.

Recommendation 2.1: Relate Agriculture and Natural Resources to General Educational Goals. *Faculty and colleges of agriculture and natural resources should provide students seeking a broad general education with curricular opportunities that reflect the rich diversity of traditional and emerging disciplines in agriculture and natural resources.*

Students need to develop a general understanding of specialized scientific knowledge as well as perspectives offered by social, historical, ethical, and philosophical disciplines. Communication abilities and aesthetic appreciation are at the heart of a general education and general public literacy. Literacy implies that individuals are equipped with both the knowledge and the cognitive skill that enable them to make rational decisions relating to their lives, adapt to change, take charge of lifelong learning and participate in the civic responsibilities of a democratic society. Both students and faculty in agriculture and natural resources are encouraged to examine and apply the values and experiences provided by general education courses. In a world which is becoming increasingly interrelated and interdependent, knowledge of other languages, cultures and modes of thinking should be strongly encouraged.

In a similar manner, agricultural and natural resource faculty and staff have a great deal to offer to other disciplines in the university system. Within the diversity of disciplines supporting traditional and emerging needs of agriculture and natural resource management is a richness of knowledge and research that will impact on the future welfare of all citizens of the globe. It is important that faculty of colleges of agriculture and natural resources in the UW System be prepared to share more broadly both knowledge of their disciplines and their insights of how the specialized knowledge interrelates with that from other disciplines to provide solutions to increasingly complex problems.

It is also essential that all students become aware of the major agricultural and natural resource issues that will impact their lives. College courses should be developed to deal with the breadth and quality of agriculture and natural resource topics for all students seeking a general education. Such courses offered by Agriculture and Natural Resources faculty, could contribute to programs of study in the humanities, social sciences, natural sciences and engineering, and provide a context for students who must become interested in and concerned about both the human and natural environments. These courses could be offered on an intercollege and/or interinstitutional basis.

In order to meet the goal of contributing to a better educated general citizenry, faculty in agriculture and natural resources should be prepared to:

1. participate in the development and offering of curricula that reflects the importance of the agricultural and natural resource disciplines in global ecology and management;
2. provide opportunities to undergraduate students interested in the social and natural sciences to have research experience with faculty;
3. provide special opportunities for undergraduates majoring in K–12 education (preservice teachers) to participate in courses or research projects that will give them an enriched background of experience from which to initiate their teaching; and
4. provide special opportunities for teachers (inservice) to participate in science curricula and research projects that will enrich the background from which they teach.

Recommendation 2.2: Pursue a Broad-Based and Dynamic Curriculum. Pursue programs of instruction in agriculture and natural resources that are broad and global in scope, contemporary in focus, up-to-date in content, and accessible to the student population generally.

The future for well-trained students of agriculture and natural resources is bright. Demand is strong in relation to current and anticipated numbers of graduates. To be competitive in the future, graduates of colleges of agriculture and natural resources will need both a broad general education in the traditional liberal arts and well-focused coursework in a major with opportunities for internships, practicums and exchange programs.

These expectations may necessitate rethinking the balance in the distribution of requirements for bachelor level training and perhaps the creation of some more broadly-focused masters programs. Multidisciplinary, cross-college and intercampus programs may need to be created and nurtured. Some institutions may try new organizational arrangements.

Consideration should be given to assessing students' knowledge base and learning skills when they enter agriculture and natural resource programs. Programs of study can then be tailored to help students develop the range of general competencies of an educated person, the characteristics desired of agricultural or natural resource professionals, and the knowledge of how to continue learning throughout one's lifetime.

Included in the curricula of colleges of agriculture and natural resources must be the study of social institutions and cultures. The curriculum should also include the study of resource development and ecosystem management, public policy development and policy implications, technology and its implications, agricultural management and marketing, human health and nutrition, and rural and urban landscapes. These topics in combination with the more traditional areas of plants and animals, soil and water, and land and capital, can provide curricular and co-curricular programs that prepare agriculture and natural resource professionals who:

1. understand and utilize the scientific approach, can evaluate information, and can create and assess alternative solutions;
2. ask value questions and appreciate a value perspective in others; and
3. are intellectually curious and have a commitment to learning as an open ended process.

Consistent with the globalization of agricultural and natural resource issues, the recognition that our social institutions must become more diverse, and the knowledge of the interdependency between current and future generations must also enable students to:

1. develop a global perspective;
2. appreciate cultural diversity;
3. understand the concept of stewardship;
4. experience cross-college courses on contemporary agricultural and natural resource issues; and
5. participate in capstone courses that bring together the natural, physical and social sciences.

In an environment of rapid scientific and technological advances and shifting issues, faculty have an absolute need for continuing education. If these ideas on how to more effectively train agriculture and natural resource students are to be implemented and tested, it will be imperative for faculty to have increased support for professional development opportunities. This support should include funding for sabbaticals, released time to attend professional meetings, and resources to support a semester or summer at another campus, a state agency, or with a private firm. In addition, the current faculty evaluation and reward system does not adequately support innovative teaching. It will need to be revamped to do so.

Recommendation 2.3: Enhance Independent Learning and Leadership. Deliver programs of instruction in agriculture and natural resources in a manner that maximizes independent learning and provide extracurricular activities that develop leadership.

A recurrent theme of agriculture and natural resource clients—students, alumni and employers—has been the importance of communication, networking, teamwork, critical thinking, technological and analytical skills. It is generally recognized that these skills are enhanced by undergraduate internships in the private sector or in local or state agencies, memberships and service as officers of student organizations, and participation in regional and national competitions in their field of study. These activities are a hallmark of UW System colleges of agriculture and natural resources. As we enter the 1990s, other opportunities need to be created, expanded and/or refined to help future agriculture and natural resource students develop independent learning skills. These include participation as:

1. junior colleagues in faculty research projects,
2. exchange students on other campuses, and
3. members of international agriculture and natural resource seminars, exchange programs or applied research projects.

Across the country, a variety of issues related to effective learning are being discussed and tested. New questions have been raised about learning environments. New educational technology is available. A consensus is emerging that multidisciplinary and multifaceted approaches to problem solving are essential, and renewed attention has been given to the central role of value questions in education. Each of these trends has implications for agriculture and natural resource education. Among these implications are:

1. Ongoing research on learning environments finds cultural variation in optimal learning styles while pointing to a need for more participatory and collaborative approaches in the classroom.
2. Telecommunications and computer technology provide opportunities for efficiently and effectively interacting directly from the classroom with universities, agribusinesses, state agencies, government and public organizations.
3. Problems of air and water quality, agricultural and agroforestry profitability and sustainability, food safety and human nutrition, food marketing and international trade, rural youth and rural elderly, to name a few, all call for specialists from a number of traditional disciplines to collaborate in research and teaching.
4. Addressing value questions in a manner that captures both the breadth of perspectives and the complexity of such questions demands clarity of analysis and the ability to simplify without becoming simplistic.

GOAL 3: SERVE A BROADER PUBLIC
FROM AN EXPANDED KNOWLEDGE BASE

Strategies. Commitment to rededicate the faculty, staff, administrators, and students to the ideals, tradition, and practice of the Wisconsin Idea.

Recommendation 3.1: Offer Integrated, Problem Solving Extension Education. Deliver holistic, balanced, problem solving education and lifelong learning opportunities to the citizenry in a timely manner.

The pursuit of excellence in extension and continuing education involves rededication to providing equality of access to knowledge to all residents utilizing all UW System faculty and academic staff including those not funded explicitly for extension purposes. To increase efficiency of faculty resources and timeliness of the knowledge extended, new technologies should be explored and evaluated to expand and improve upon information delivery methods.

To responsibly transfer knowledge to the citizens of the state, extension programs must pursue excellence by:

1. integrating strategies directed at improving profitability and marketing with health, environmental stewardship and socio-economic viability of communities;
2. helping individuals and communities focus on long-term implications of all decisions rather than only on short-term expediencies;
3. providing problem solving education in a broadly encompassing context to help individuals and communities understand how

knowledge contributes to the complete person and to a sustainable community;

4. articulating and transferring research-based knowledge to agriculture and natural resource clients directly or to information wholesalers in the private sector who transfer information to clients;
5. facilitating alternative venues for the delivery of lifelong learning opportunities; and
6. defining an incentive and reward system that makes clear to all faculty the expectation for public service.

American higher education emphasizes excellence with utility. In this model, science is applied to topics of relevance to society. Over time, the range of topics considered relevant for scientific research expands. Consequently, the extension education curriculum must also be broadened.

But the extension system faces some great challenges. The information needs of Wisconsin's nearly five million citizens and their complex businesses are both wide ranging and sophisticated. For example, addressing resource management issues and local government education needs require county extension faculty with broad and balanced training; but, the generalist agricultural extension agent often can't meet the needs of the advanced, specialized producer or business person. Greater specialization at the county level is desirable. The limited numbers of extension faculty, even when augmented by the outreach activity of research and instructional faculty, cannot adequately interact with the millions of potential information "consumers" through traditional approaches. New approaches and new communication technologies are needed urgently.

Areas of concern to the extension curriculum include agri-business management, natural resource protection and management as well as natural resource development, public policy, rural family and community issues, rural revitalization, global issues affecting agricultural markets and rural peoples, nutrition, food safety, risk assessment, biotechnology and genetically-altered organisms. The curriculum needs to give special emphasis to:

1. understanding scientific approaches and outcomes,
2. developing critical thinking and analytical skills,
3. facilitating informed participation in public policy debates and the management of public resources, and
4. developing methods to effectively communicate research findings to a diverse public and incorporating client feedback to refine those methods.

Natural resource management and food safety are two current high priority extension issues that illustrate the relevance of these guidelines. Helping the citizenry understand the changing demographics and changing preferences in the use of natural resources, and the economic importance of natural resources and the development of their value-added products can be vital to the state's economy. A more complex issue is policy education for publicly owned natural resources. In the food area, the immediate task is to provide risk assessment and risk management education to deal with chemicals in the environment and the promotion/regulation of other technologies.

To serve a broader public and to maintain and enhance the quality of that service, UW System programs in agriculture and natural resources will need to:

1. establish ways to systematically involve citizens and organizations, public officials and businesses in their program planning;
2. use modern communication and information transfer technology to assure full utilization of and access to faculty expertise;
3. expand programs directed at professionals and private service delivery networks to maximize multiplier effects;
4. ensure that Wisconsin citizens receive the full benefit from functionally integrated and coordinated extension and vocational technical programs;
5. provide funds and encouragement for increased educational opportunities and sabbaticals to keep faculty on the cutting edge of research and technological, environmental, governmental, regulatory, and social issues; and
6. bridge local information needs with broadly trained county extension agents in cooperation with specialists with regional responsibility who share their expertise across clusters of counties.

Like the instructional curriculum, the extension curriculum depends on research, and, if working properly, helps guide attention to new research needs. Public education generates new research questions, reveals a need for new data bases, and presents new techniques requiring refinement and evaluation.

Questions identified by the public, may, in turn, necessitate the development of alliances with other parts of the university and other state agencies, such as the medical schools on nutritional and health issues; natural resource and agriculture departments on farm chemicals and water quality issues; and business schools or recreation and tourism departments on rural development planning.

GOAL 4: PROVIDE INTEGRATED, HOLISTIC, AND ANTICIPATORY EDUCATIONAL SERVICE TO WISCONSIN

Strategies. Commitment to ensure that the UW System is responsive to macro issues and nurtures those scholars capable and willing to transcend disciplinary boundaries and address these issues.

Recommendation 4.1: Create Mechanisms to Address Comprehensive Issues. Establish mechanisms to bring together scholars whose primary interest is to identify and address major agriculture and natural resource issues in an interdisciplinary and interinstitutional way.

Departmental and disciplinary loyalty have historically inhibited interdisciplinary research, instruction, and outreach. Departments are often protective of their perceived intellectual domains. Graduate school strongly socializes new faculty in narrow specialities within a single discipline and promotion, tenure, and salary increases are determined by peers in the department or disciplinary groups. Typically, departments place the highest value on contributions to the discipline.

In addition to disciplinary pressures, availability of federal research fund tend to set the research agenda. Researchers naturally respond to the availability of funds, and often assume that the sponsor's priorities reflect societal need. Local or state needs may not be addressed by the traditional research funding institutions and agencies.

The result of these combined pressures is highly focused research that often serves the profession or branch of science more than the state or society generally.

Thus, there is a need to create mechanisms to identify and address broad issues in an interdisciplinary and interinstitutional way. Efforts must be made to find ways to engage faculty, staff, and students from a single campus and/or from a number of campuses in interdisciplinary study of important, emerging issues. Sources of funding for these interdisciplinary efforts must be developed for the studies to be undertaken.

Recommendation 4.2: Formulate Reward and Evaluation System for Interdisciplinary Faculty. Develop and implement appropriate standards for rewarding interdisciplinary participation and productivity.

Faculty who practice interdisciplinary approaches receive less support from traditional disciplines. Not only do interdisciplinary faculty need a supportive structure, they also must be evaluated and rewarded on the basis of their interdisciplinary contributions. Intrinsic rewards that come from the challenge of working with complex systems and the knowledge of providing direct service to the citizens are not enough. The challenge is to create structures and mechanisms for recognizing and rewarding excellence, and to assure a secure environment in which to develop interdisciplinary approaches to intellectual understanding and problem solving. It must be remembered, however, that interdisciplinary efforts should not come at the expense of dismantling disciplinary competence. Interdisciplinary efforts can make their contributions to scientific understanding only if there is a strong disciplinary base.

Evaluation of interdisciplinary research, instruction, and outreach remains elusive. Resource issues span the disciplines and their complexity demands interdisciplinary work and its evaluation. Consequently, new standards of assessment must be developed.

Recommendation 4.3: Build upon Past Interdisciplinary Experience. Review operational style and contributions to knowledge and public service of successful interdisciplinary centers, and incorporate the knowledge gained into the organization and operation of evolving interdisciplinary centers.

In an attempt to be responsive to major issues, multidisciplinary centers have been created. Indeed, interdisciplinary research, instruction, and extension programs have a long history with mixed success. A barrier to success has been the failure to develop the scope and productivity expected because:

1. the long-term financial future of interdisciplinary centers often remain uncertain and without strong institutional backing;
2. faculty and staff view the problem from several multidisciplinary perspectives but do not evolve into a team that blend disciplinary perspectives into an interdisciplinary approach;
3. the centers provide no long term security and status; or
4. long-term missions for the centers are not articulated.

Since the UW System has several interdisciplinary centers based on alternative institutional structures and reward systems, a thoughtful review of successful centers would strengthen frameworks for future interdisciplinary activities.

GOAL 5: STRENGTHEN AND AUGMENT COLLABORATIVE APPROACHES

Strategies. Commitment to the development of strong linkages with agriculture and natural resource programs at public universities in neighboring states, among colleges, schools, and programs of different UW System institutions, and with state agencies and the private sector.

Recommendation 5.1: Expand Regional Cooperation. Establish a network of upper midwest public institutions with programs in agriculture and natural resources in order to improve research, extension, and instructional capacity.

We face the beginning of a new century with increased competition for limited resources. The diversity, complexity and magnitude of issues confronting agriculture and natural resources require that higher education make choices among areas of expertise in order to emphasize, build, and maintain appropriate institutional infrastructure and faculty strength.

If single states can no longer afford to maintain both breadth and depth in all the principal scientific areas that agriculture and natural resources require, regional approaches are an alternative. Indeed, collaboration in instruction, research, and extension may be essential to the maintenance of a critical mass of experts able to engage in interdisciplinary problem solving on a range of agricultural commodities and natural resources questions as well as on basic and applied regional agriculture and natural resource issues. In any event, the pressure for identification and funding of priority areas will become increasingly urgent.

A network of regional cooperation can be built on some successful pilot projects. For example, several midwest universities have joined forces to establish and support satellite communications among campuses. Such joint sponsorship will allow collaborative instructional programs to be developed at and offered from one campus and received by another. Wisconsin's county extension offices are connected via satellite downlinks to land grant and other campuses across the nation. Intercampus and interstate instruction and extension programs will be facilitated by telecommunication technology. For nearly 30 years, six land grant universities in the midwest have worked together on international agricultural projects through the Midwest Universities Consortium on International Activities (MUCIA).

In addition, collaborative interstate research projects are underway. Maintaining excellence in science is costly, and costs are soaring. Research projects that involve faculty from various universities can control long-term costs while developing a critical mass of expertise that can be tapped to meet regional needs. Linkage of the agricultural experiment stations of two or more North Central region institutions has been one effective way to bring research expertise from several universities to bear on common problems.

At present, 168 north central regional research and extension committees exist and 60 cooperative projects are underway. To facilitate further regional specialization and cooperation, the UW System should formalize relations with public universities in Iowa, Illinois, Michigan, and Minnesota offering degree granting programs in agriculture and natural resources. Appointment of a 5–8 person committee to establish mechanisms for implementation of regional cooperation in research, instruction, and extension is needed. Top priority should be given to:

1. increasing areas of research collaboration and complementarity, and
2. initiating and expanding exchange and, possibly, sharing of instructional, research, and extension faculty and information.

Recommendation 5.2: Strengthen Interinstitutional Cooperation Among UW System Campuses. Strengthen and expand UW System collaborative arrangements for research in agriculture and natural resources and its application through extension.

In order to meet the need to more efficiently bring together the critical mass of agricultural and natural resource expertise in the UW System, the UW System Board of Regents established the UW System Consortium for Extension and Research in Agriculture and Natural Resources in 1976. This consortium has provided an opportunity for faculty from four UW System institutions to work collaboratively, connecting applied research important to agriculture and natural resources with basic science and other faculty from UW–Madison. The arrangement has strengthened the interactive basic-applied science continuum, underwritten over 100 applied research projects, benefited classroom teaching, provided support for graduate student research, and helped launch a rural leadership program. The Consortium provides a strong basis for collaboration and has evolved a workable structure and process.

However since the Consortium was established, the scope of the issues has expanded. Competing demands for state and federal resources and an exceptionally strong academic market spurred by large numbers of faculty reaching retirement, produce new demands for innovative ways to organize new combinations of experts. While the research mission, resources and expertise on these issues is concentrated in doctoral-granting institutions of the UW System, research expertise within the UW System is certainty not limited to those institutions. To efficiently tap the creative energy and expertise needed to study a broad set of basic and applied research questions in agriculture and natural resources, consideration should be given to the kind of boundaries that should be placed on those who can participate in that research.

To meet the recognized expanding need for broad, comprehensive, problem-focused, interdisciplinary research efforts, new programs should be developed to include projects that bring together the expertise of social, biological and physical scientists around current issues. For example, during the decade of the 1990s, acid rain, global climate change, air and water quality, biotechnology, production efficiency, nutrition and food safety, multiple use of resources including tourism, ecosystem management, forest and farm sustainability, and restoration of altered ecosystems are some of the topics that are of major concern.

Interinstitutional programs in agriculture and natural resources should:

1. identify high priority issues and problem areas;
2. attract appropriate faculty into research addressing these areas; and
3. provide coordination of these research projects by linking investigators and avoiding duplication.

To augment learning and teaching opportunities for faculty, staff and students working in these new projects, the programs should create a variety of activities such as:

1. faculty, staff and student intercampus exchanges,
2. faculty exchanges that provide professional development opportunities;
3. projects that develop new instructional materials based on emerging scientific knowledge; and
4. projects with defined opportunities for graduate students to participate in exchanges.

Recommendation 5.3: Cooperate with State Agencies and the Private Sector. Improve mechanisms that encourage and expand joint policy development, research and application programs with state agencies and the private sector on a sustained basis.

The primary goals of the university are to 1) educate the citizens and professionals of tomorrow, 2) pursue scientific inquiry furthering the foundation of knowledge and 3) disseminate that knowledge to the general public and specific user groups. The private sector and state agencies apply knowledge to problem-solving, policy formulation and product development.

The university and state government have often teamed with the private sector to apply knowledge to agriculture and natural resource problems. Prime examples include reforestation of northern Wisconsin, re-establishment of fisheries in the Great Lakes, prairie restoration along highway corridors and the establishment of a Center for Integrated Agricultural Systems to study sustainable agriculture. Continued pursuit of knowledge and its transfer to problem solving situations through state agencies merits further effort. Actions thus far have been sporadic rather than sustained. Mechanisms should be explored by individual educators and state agencies that would facilitate sabbatical-type opportunities such as:

1. exchange of personnel on short-term assignments;
2. affiliate appointments for state agency personnel; and
3. a scholarship program for state agency researchers and management staff to pursue advanced education and training while at the same time contributing to instruction, research and extension programs.

It is time to forge joint programs of more lasting benefit to Wisconsin's citizens.

Mutually beneficial linkages between academic scientists and the private sector should be encouraged. While it is necessary to protect the professional integrity of all parties involved, closer working relationships would enhance faster transfer of knowledge to problem-solving, provide forums for joint experiments, and further the integration of the educational system with the management of agriculture and natural resource systems. Arrangements ought to be examined for cooperatively involving private sector specialists and scientists in instruction, research and extension programs of the UW System.

GOAL 6: DIVERSITY BY DESIGN

Strategies. Commitment to recruit and retain women and minority faculty, staff, administrators, and students in colleges of agriculture and natural resources.

Recommendation 6.1: Attract and Retain Women and Minority Faculty, Staff, and Administrators. Diversify the human resources of the colleges of agriculture and natural resources by giving high priority to recruitment of women and minorities for faculty, staff, and administrative positions.

Women and minorities historically, and even now, are significantly under-represented in the agriculture and natural resource professions, particularly in production agriculture and virtually all natural resource areas. While the particular composition of agriculture and natural resource professionals does not reflect the diverse nature of the people who have been engaged in agriculture, it does mirror a history of programs targeted along gender lines.

With time, the traditional sources of agriculture and natural resource professionals have declined relatively and absolutely. No longer are agricultural or rural people the exclusive caretakers of the rural landscape.

Consequently, the old patterns of faculty, staff, and administrative recruitment for colleges of agriculture and natural resources should and must change. Of necessity, colleges of agriculture and natural resources must now seek faculty, staff, and administrators beyond the bounds of the traditional pool. This, in turn, means reassessing recruitment policies and procedures and conditions for retention,.most notably the professional climate for a heterogeneous faculty.

For the future of agriculture and natural resource programs, it is imperative that their teachers, researchers, and leaders begin to reflect the diversity extant in the society generally and in agriculture and rural areas specifically. Moving quickly to promote minorities and women to faculty and administrative positions will be important to the maintenance of both the quantity and quality of teachers and researchers in agriculture and natural resources in the future. The addition of women and minorities can provide diversity in research perspective and in the frameworks for addressing research questions. Faculty and administrative diversity is also vital for recruitment and retention of a diverse student body.

Recommendation 6.2: Recruit and Retain a Diverse Student Body. Engage in special efforts to attract to programs of agriculture and natural resources minorities and women from a broad range of high school backgrounds and from the returning adult population, and in addition, males from metropolitan areas.

Consistent with the situation nationally, demand for agriculture and natural resource professionals is strong relative to the supply of graduates from colleges of agriculture and natural resources. Employers in the food, fiber, and forest industries, have and must turn to graduates of other programs to fill their labor force needs. Higher enrollments as recently as the late 1970s suggest that the most important limiting factor is interested students rather than capacity.

The traditional source of agriculture and natural resource students need not and should not be abandoned. However, with a decline in the number of young, white males from a farm background, it is imperative that colleges of agriculture and natural resources actively seek students from other gender, ethnic, and residential backgrounds and of a non-traditional age. Meeting the rapidly developing shortage of trained professionals necessitates bold action.

Special programs to identify women and minorities from urban and rural areas along with urban males should already be in formative or experimental stages. The scope of targeting potential students should be as broad as the modern concepts of agriculture and natural resources. Urban high school and technical programs with strong agricultural and environmental curricula will be important new allies in identifying interested and talented students. For recruitment and retention, the importance of women and minority faculty and staff as role models must be recognized.

At the same time, alliances should be strengthened with the UW Centers to encourage interested students to pursue programs in colleges of agriculture and natural resources. Alliances should also be promoted with elementary and secondary schools to familiarize students with the diversity of career opportunities in agriculture and natural resource fields.

Recruitment and retention of students, including women and minorities from urban and rural areas, depend on well-targeted recruitment and retention programs; high quality curriculum creatively delivered by a diverse faculty and staff; and periodic assessment of the array of academic programs needed to meet demands for a changing set of specialties.

GOAL 7: IMPLEMENTATION, FLEXIBILITY, AND PLANNING

Strategies. Commitment to study change in resource systems and adapt programs to a dynamic and uncertain world.

Recommendation 7.1: Evaluate Progress and Monitor Change. Encourage the UW System Board of Regents to request a report on the progress made by the institutions in implementation of the recommendations contained herein.

The time for action in agriculture and natural resource education is now. The world is changing and our approach to education must likewise change through an on-going, evolutionary process. To be proactive and anticipate the future and to monitor changes recommended in this report in relation to evolving trends, the Board of Regents should request periodic reports on progress being made in the implementation of committee recommendations. A review process should be established to regularly evaluate the agriculture/natural resource programs; and administrative structures should be created to facilitate interdisciplinary and inter-campus curriculum, instruction, extension and research. In addition, a periodic and systematic evaluation of the structure and framework of the educational delivery system for agriculture and natural resources should be conducted to ensure that the UW System continues to make efficient and effective use of its human and financial resources.

The UW System Steering Committee for Strategic Planning for Agriculture and Natural Resources has invested considerable energy in and provided collective wisdom for the future of agriculture and natural resource programming. Too often, intense effort is invested in the preparation of a report and not in implementation and progress. The Committee desires to impress upon the UW System the urgency for action and the responsibility to monitor change in agriculture and natural resource education.

NOTES

This is our strategy for leadership and a blueprint for Wisconsin's tomorrow. It is the belief of those who participated in this unique strategic planning exercise, leaders from agriculture and forestry business communities, non-profit and environmental organizations, and the academic community, that the translation from the Wisconsin condition can be easily made. Every state with an agricultural and natural resource program needs to carefully examine alternative futures in research, instruction and extension. Just as Wisconsin created a picture of its local and regional responsibilities in agriculture and natural resources by comparing it to the national and international scene, other states can utilize these findings as an outline for developing their own strategies for the future.

BIBLIOGRAPHY

Achieving Faculty Diversity (1988). Madison, WI: University of Wisconsin System.

Agriculture and Natural Resources: Education for 2020, Conference Program and Consultants Papers (1989). Madison, WI, University of Wisconsin System.

Agriculture and Rural Viability (1988). Raleigh, NC: North Carolina State University.

Agricultural Research Service Program Plan. (1983). U.S. Department of Agriculture: Miscellaneous Publication Number 1429.

Agricultural Research Service Program Plan. (1985). U.S. Department of Agriculture: 6-Year Implementation Plan, 1986–1992.

Alternative Agriculture (1989). Board on Agriculture, National Research Council. Washington, D.C.: National Academy Press.

Aquaculture Program. Madison, WI: University of Wisconsin–Madison.

Brundtland, Gro Harlem, et al. (1987). *Our Common Future: The Report of the World Commission on Environment and Development.* New York: United Nations World Commission on Environment and Development.

Cooperative Extension System National Initiatives. (1988). U.S. Department of Agriculture Extension Service.

The Cooperative Extension System (1988). U.S. Department of Agriculture Extension Service.

Creswell, John W., et al. (1990). *The Academic Chairperson's Handbook.* Lincoln, Nebraska: University of Nebraska Press.

Curricular Innovation for 2005 (1987). Madison, WI: University of Wisconsin–Madison.

Design for the Future (1987). Madison, WI: University of Wisconsin–Extension— Cooperative Extension Service.

Ensign, David E (1988). *Emerging Issues.* "Prodigal Crops: A Review of Proposals for Sustaining Agriculture." Lombard, IL: Midwestern Legislative Conference.

Extension in Transition: Bridging the Gap Between Vision and Reality. Report of the Futures Task Force to the Extension Committee on Organization and Policy (1987). Virginia: Virginia State University.

Field, Donald R., and William R. Burch, Jr. (1988). *Rural Sociology and the Environment.* Westport, CT: Greenwood Press, Inc.

Focus on the Future: Executive Summary (1988). College Station, TX: Texas A&M University.

Focus on the Future: Options in Developing a New National Rural Policy (1988). College Station, TX: Texas A&M University.

Forest Management and Economic Development: New Directions for Wisconsin (1977). Wisconsin: The Forest Industries Development Subcommittee of the Economic Development Coordinating Committee.

Forestry Facts (November 1987), Nos. 29, 30, 31, 32, 34, 35.

Forestry Research: A Mandate for Change (1990). Committee on Forestry Research, National Research Council. Washington, D.C.: National Academy Press.

Future of the Dairy Industry in Wisconsin: Serious Challenges, Tremendous Potential. (1987). Platteville, WI: Wisconsin Dairy Task Force 1995.

Giese, Ronald L. (1988) "Forestry Research: An Imperiled System." *Journal of Forestry*, 86, No. 6 (1988), 15–22.

Giese, Ronald L, A. Jeff Martin, and Jeffrey C. Stier (1985). *Forestry Research Priorities for Wisconsin: Perceptions of User Groups.* Madison, WI: University of Wisconsin–Madison.

Governor's Council on Forest Productivity. (1982). Madison, WI: State of Wisconsin.

Haney, Alan, and Rick Wilke (1989). *Opportunities: Agenda for the '90's.* Stevens Point, WI: University of Wisconsin–Stevens Point.

Implementation of the New Directions for the Cooperative Extension System. (1988). ECOP Subcommittee.

Lindberg, Richard D., and H. James Hovind (1983). *A Strategic Plan for Wisconsin's Forests.* Wisconsin: Department of Natural Resources/Bureau of Forestry.

Lindberg, Richard D., and H. James Hovind (1985). *Wisconsin's Forests—An Assessment.* Wisconsin: Department of Natural Resources/Bureau of Forestry.

Merz, Robert W. (1987). *Forest Atlas of the Midwest.* Forest Service, U.S. Department of Agriculture.

National Enrollment Workshop for Colleges and Schools of Agriculture (1988). Columbus, OH: The Ohio State University.

National Research Council (1989). *Investing in Research.* Board on Agriculture, Washington, D.C.: National Academy Press.

Operation Change: Developing Human Capital to Secure American Agriculture (1988). National Association of State Universities and Land-Grant Colleges.

Proceedings National Curriculum Revitalization 2005 Conference (1987). Manhattan, KS: Kansas State University.

Raile, Gerhard K. (1985). *Wisconsin Forest Statistics, 1983.* Resource Bulletin NC-4. St. Paul, MN: Forest Service—U.S. Department of Agriculture.

Research Initiatives. (1988). A Midterm Update of the Research Agenda for the State Agricultural Experiment Stations. College Station, TX: Texas A&M University System.

Schaller, Neill (1987). *Social Science Agricultural Agenda Project: Proceedings of Phase I Workshop.* New York: U.S. Department of Agriculture.

Self Study for the North Central Association of Colleges and Schools (1988). Vol. II. Madison, WI: University of Wisconsin.

Status of Wisconsin Farming, 1988. Madison, WI: University of Wisconsin–Madison & University of Wisconsin–Extension.

Status of Wisconsin Farming, 1989. Madison, WI: University of Wisconsin–Madison & University of Wisconsin–Extension.

Strategic Planning: Issue Identification and Development for the Cooperative Extension System (1988). U.S. Department of Agriculture Extension Service.

Understanding Agriculture (1988). Washington, D.C.: National Academy Press.

University of Wisconsin System Planning Task Force on Agriculture. (1974). Madison, WI: University of Wisconsin System.

Upper Great Lakes Region Atlas (1979). Washington, D.C.: Upper Great Lakes Regional Commission.

Wisconsin Natural Resources 8, No. 2 (March–April 1984).

Wisconsin Forest Productivity Report (1979). Washington, D.C.: Forest Industries Council.

Wisconsin's Forest Resources: Present and Potential Uses. (1976). Madison, WI: University of Wisconsin–Madison Research Bulletin R2844.

The University of Wisconsin System Steering Committee for Strategic Planning for Agriculture and Natural Resources

Stephen M. Born, UW–Madison and UW–Extension
John E. Cottingham, UW–Platteville and UW–Extension
Wilmer Dahl, Wisconsin Milk Marketing Board
Terry L. Ferriss, UW–River Falls
Donald R. Field, UW–Madison, Chair
Edwin M. Foster, UW–Madison
Fritz Friday, Friday Canning Corporation, New Richmond
Donald Haldeman, Wisconsin Farm Bureau
Lowell L. Klessig, UW–Stevens Point and UW–Extension
Patrick Luby, Oscar Mayer Foods Corporation, Madison
Rod Nilsestuen, Wisconsin Federation of Cooperatives
Thomas H. Schmidt, Wisconsin Paper Council
Ayse C. Somersan, UW–Extension and UW–Madison
Caryl Terrell, Sierra Club
Richard H. Vilstrup, UW–Madison
Paul H. Williams, UW–Madison
Thomas M. Yuill, UW–Madison

ADMINISTRATIVE STAFF TO STEERING COMMITTEE
Wava G. Haney, UW System Administration
Gorden Hedahl, UW System Administration
Lynn Paulson, UW System Administration
Ruth Carlson Robertson, UW System Administration

About the Editors and Contributors

Duane Acker is president emeritus and distinguished professor of animal science and industry at Kansas State University.

Charles M. Benbrook, of Benbrook Consulting Services, is the former executive director of the board on agriculture at the National Research Council of the National Academy of Sciences.

Donald R. Field is associate dean of the College of Agricultural and Life Sciences and director of the School of Natural Resources at the University of Wisconsin–Madison.

John C. Gordon is dean of the School of Forestry and Environmental Studies at Yale University.

Wava G. Haney is professor and chair of the Department of Anthropology and Sociology at the University of Wisconsin Centers and former academic planner of the University of Wisconsin System.

Mary E. Heltsley is dean of the School of Home Economics at the University of Minnesota.

Mark B. Lapping is dean of the School of Planning and Public Policy at Rutgers University and former professor of regional and community planning at Kansas State University.

Russell W. Peterson is vice chair of the Better World Society, president of the International Council for Bird Preservation, president emeritus of the National Audubon Society, former chair of the President's Council on Environmental Quality, and former governor of Delaware.

Ruth Carlson Robertson is associate vice chancellor for planning and accountability of the University of Maryland System and former acting assistant vice president for academic affairs of the University of Wisconsin System.

Kenneth A. Shaw is president of the University of Wisconsin System.

Eugene P. Trani is president of the Virginia Commonwealth University and former vice president for academic affairs of the University of Wisconsin System.

M. Ann Tutwiler is the associate director of the International Policy Council on Agriculture and Trade and policy associate for resources for the future at the National Center for Food and Agricultural Policy.

James J. Zuiches is director of the Agricultural Research Center at Washington State University.

About the Editors and Contributors